P9-BII-336

So Much, So Fast, So Little Time

So Much, So Fast, So Little Time

COMING TO TERMS WITH RAPID CHANGE AND ITS CONSEQUENCES

Michael St. Clair

AN IMPRINT OF ABC-CLIO, LLC
Santa Barbara, California • Denver, Colorado • Oxford, England

Library of Congress Cataloging-in-Publication Data

St. Clair, Michael, 1940–
So much, so fast, so little time : coming to terms with rapid change and its consequences / Michael St. Clair.
p. cm.
Includes bibliographical references and index.
ISBN 978-0-313-39275-7 (alk. paper) — ISBN 978-0-313-39276-4 (ebook)
1. Change. 2. Change (Psychology) 3. Technological innovations—Social aspects. I. Title.
BD373.S7 2011
155.2'4—dc22 2011008506

ISBN: 978-0-313-39275-7
EISBN: 978-0-313-39276-4

15 14 13 12 11 1 2 3 4 5

This book is also available on the World Wide Web as an eBook.
Visit www.abc-clio.com for details.

Praeger
An Imprint of ABC-CLIO, LLC

ABC-CLIO, LLC
130 Cremona Drive, P.O. Box 1911
Santa Barbara, California 93116–1911

This book is printed on acid-free paper ∞
Manufactured in the United States of America

for Madison and Sadie
who will have to live with these changes

Contents

Introduction

What is happening around us? Can we see the events of daily life with new eyes? Can we gain insight into why things are suddenly so very different from what they were?

This book looks at *the* characteristic experience of modern life: change, massive and broad, occurring at dazzling speed.

Changes show themselves wherever we look: in families and relationships, how we travel and how we earn our living, how we communicate and entertain ourselves, how we deal with money and handle our anxieties. This book aims to find words and categories for making sense of this swirling change. Most readers will share some experiences explained in this book.

I explore many of the events, trends, and patterns that make up the big changes that we all have encountered. An overview is offered in an effort to see the forest amidst the trees, and to find the meaning in chaotic shifts and transitions present in contemporary life. I notch some benchmarks to track the pace of change with its variable tempo—headlong here, lagging there. The causes of change are multiple; rates of change variable.

A second goal of this book is to examine how the changes occurring around us are changing us.

The most powerful engines of change today are the computer and related digital technologies, especially the Internet. While this technology overshadows and influences all others, a stream of ever-new models of cell phones and handheld devices enables us to communicate and exchange information to a degree undreamed of before.

With but an occasional lapse, I strive for neutrality in judgment as I ponder the shifts and whirlwinds of change in everyday life. For the most part, my delight and admiration for many modern marvels will be clear, but I have found in interviews and research that not everyone appreciates change. I met with a range of responses: some people are exhilarated with new possibilities; others want to close their eyes and block their ears because they feel the world is racing too fast, and has become too turbulent.

Rapid change has consequences. We are not like our parents, nor—gasp—our grandparents; our children are not like us. Look at the average teenager to see how the future has arrived with new rules and magical technology. More than in any other period in history, children are very different from their parents. Furthermore, people live in a turbulent and information-drenched environment where their expertise lasts for only a brief period. "What? You're changing our computer system? I haven't even gotten used to the one we have now."

The expansion of possibilities of all types creates a more complex world, with consequent confusion about conventions and ways of thinking. Some people react with uneasiness, uncertainty, and a feeling of loss of control over their lives. Some older adults wonder where the world they grew up in has gone.

Why bother to scrutinize flux and change? What are the consequences of all these current changes that touch all of us? We live in extraordinary times, and I propose that to better understand the current nature of life in America enables us to live better, more engaged in the present. Most importantly, change changes us. We think and respond differently. Over time, and in countless small steps, we become different—not necessarily worse or better. Just different. To peer into the heart of change is to have insight into why things are the way they are and how to adapt better.

Acknowledgments

I am very grateful to the colleagues, experts, and friends whom I consulted and who read parts of this book in draft. I am especially grateful to Debbie Carvalko, Sarah Allen Benton, and Richard Griffin. I so appreciate the support and help from Alan Wolfe, Frank Scully, Andrew Brown, Padraic O'Hare, Fran Grossman, Claire Lang, Travis St. Clair, Kimberly Smirles,

Susan Keane, Jennifer Nepper Fiebig, Lawrence Wangh, Sandy Robbins, Jocelyn Hoy, Fredrick Abernathy, Thomas Schnauber, Karen Hogesin, Nino Darrico, and Paul Crowley. And lastly, I am grateful to my wife, Roslin Moore, for her patience and long-suffering while I labored away in my cave.

1

What Is Happening to Us? And Why?

"Oh, No! Not More Change!"

It's Monday morning and I can feel the anxiety and apprehension. More new stuff to manage.

I'm slumped in my seat at a faculty meeting during which the dean, my boss, keeps announcing new policies and changes—the new software package, shuffling of office space, new instructions on how to help students write better, and, oh yes, the new phone system arrives this week. "Any questions?" he asks. I've scarcely gotten used to my old "new" computer. As I hurry back to my office, I have a stack of e-mails to answer, my cell phone is vibrating, and my answering machine is blinking with new messages.

Life races along in my workplace. The new has become normal—and faster than ever. Change is the one constant.

Recent experiences cry out how rapidly changes appear, not just at my work place, a thriving college in Boston, but also in virtually all other areas of life. Transformations occur in my traditional profession of college professor, and even more so in other occupations, be it manufacturing or health care or the computer and software industries. Economic turning points, political changes, and ecological crises add complexity and uncertainty to our shifting culture.

Who Is Changed? Who Is *Not* Touched by Change?

Change affects people to varying degrees. As I looked around me at my colleagues at that faculty meeting I wondered if they were as uneasy—even distressed—at the range and speed of changes. Some are young and members of the "Google generation," digital natives who have grown up amidst the information revolution with computers familiarly always at hand. Later in the same day, however, a middle-aged clerk at Home Depot mentioned to me as she used one of the portable scanners: "Things change so fast I can't keep up." I heard a financial consultant mutter, "I don't like change." A middle-aged member of a yoga class I am in told me how much she worries about her children and grandchildren because of troubling social changes.

Not long ago my wife and I went to an Apple store to upgrade our cell phones. The youthful clerk, bubbling with enthusiasm, put several gleaming new smart phones on the counter and rhapsodized about the many features and "apps" of each model and prattled on and on about the complex service plans available. Most of the phones were capable of amazing functions, able to tap into the Internet's infinite resources, and offering more functions than I would ever use. I left the store feeling awed and a bit overwhelmed and out of it—or, at least, falling behind the times. It was yet one more new and exciting technology available at an affordable price, a technology with such a wide range of features I would probably be able to exploit only a select few. Nevertheless, in a few months, an improved and even better model would appear. Any one of these smart phones holds more computing power and speed than desktop computers of just a few years ago and so much more than the primitive computers that guided astronauts to a landing on the moon.

But the experiences that I describe in this book are not just with telephone features or workplace software. The communication and information revolution, characterized by smart phones and the Internet, has not grown sluggish. If anything, it is accelerating, as advanced technologies continue to spin off innovative changes like sparks from a fiery pinwheel. Will I ever get ahead of the curve enough to catch my breath? As soon as I master some software or an iPad or a new phone, something newer bumps me out of my comfort zone. The rapid, sometimes troubling, sometimes marvelous changes of contemporary life touch all of us in some way.

Unprecedented Change

The changes are not just in innovative electronic devices but also the new behaviors and new ways of thinking that the information revolution generates. I can keep track of phones and computers, because they are objects I can see and touch. But the new social demands, shifting rules, and accelerating time that accompany these technologies present challenges. These accompanying changes, side effects of technological innovations, psychologically make greater demands. These unintended consequences or side effects compel us to keep up a rapid pace, to alter our behavior and our mental responses. Further, media, whether television, radio, or the Internet, gush higher and higher waves of data and information.

The experience of change is *the* modern experience: a shapeless sense of urgency, of not enough time, a vague sense of anxiety and uncertainty, of more and more, faster and faster. If we are at all engaged in American culture, we feel these changes pressing on us. *Something unusual and special is happening now.* The rising curve of change is more steep than ever. Indeed, we are living in a world of change, and change itself seems to be the most permanent feature of contemporary life. Change is dizzying, sometimes exhilarating, often disorienting. Change is the only constant.

I hasten to add that these innovations and novelties, the steady appearance of the new and the different, *are changing us*. Even if I fled to a hut in the woods or became a monk in an isolated monastery seeking a stable and traditional life, I would meet change and be altered, in turn, by change. Those who are unemployed or homeless or at the margin of society, willingly or not, cannot escape change and the new acceleration, even if they don't experience it to the same degree as the contemporary middle-class worker.

Change is happening now, today, here, as you read this: change that is steady, unrelenting, sometimes tiny, sometimes large, commonly around us, but most importantly, happening to us. The changes can be trivial and negligible, such as a new kind of latte at Starbucks, or massive and intrusive, such as all the security measures following since 9/11. Our reactions and internal changes are more elusive, but more urgent, because they involve our identity and our very being in the world. They involve how we organize our minds, how we think, and how we react.

What Is Change?

That which changes must have some continuity; otherwise the object or behavior that is changing has become something totally different. We recognize our mother's face when we are children. Her face is different when we are teens or return home from college after not seeing her for months or years—yet she is our mother, though she has changed. A hammer is a tool *for hitting* and has evolved different shapes that suit a new function—it is not a *grasping tool* like pliers, which are very different.

When a behavior changes, it still retains some familiarity. Ballroom dancing involves many movements and innovative new styles, but the stable element is motion and the presence of a dance partner. Playing basketball, by contrast, is a totally different kind of movement than dancing; it too involves moving and others are involved, but it is not dancing. When a tool or a behavior changes, it is not so totally new or different, otherwise it becomes a different tool or behavior.

New technologies tend to be extensions of human bodies and senses. A hammer is a more effective tool for hitting than our fist, a telescope sees farther than our eye, a computer can remember more and compute faster than our minds. Now technologies are changing very quickly and this has significant implications for our human minds and actions.

To capture accurately a picture of change, we need to compare *before* and *after*. We know someone has lost weight if we remember that person with the extra pounds. To say our life has sped up, we need to recall how life once was slower and less hectic. What are the markers, then, of particular changes?

Here are seven areas of change with markers that help us gauge the extent and pace of change. Later chapters will explore each area in more depth.

Seven Areas of Change

1. Information and the Internet

The information revolution, a gush of data and facts, stands on two legs—the computer and the Internet—which make possible the gathering of vast amounts of information in digital form, storing it and communicating it rapidly. Neither of these tools—computers and the Internet—burst upon the scene without prior technical innovations, such as the silicon chip and fiber-optic cables; but the ability to manage information quickly and easily has so altered our lives that we can no longer appraise our lives without them. For instance, I just came back from my health club where an attendant at the front desk checked me in by waving my plastic membership card in front of a scanner, which entered my number into a computer that recognized me and flashed my image up on her screen. I had gone to the library before that and looked up several books on the library's computerized catalogue. Long gone are those drawers of paper cards used in libraries to track books. Now the benefits and presence of computers have become so commonplace that we have grown accustomed to them and scarcely notice their presence.

Virtually everything I do is connected in some way to computers and the Internet. I had gone to the library because I had received an e-mail notifying me that the library was holding a book I reserved. Before heading for the health club I stopped at the bank and used the ATM, which is networked by computer to my bank and my account. I stopped at a Costco store and the cashier scanned each of my purchases so that inventory could be tracked and my monthly bill properly calculated. When I arrived home I checked both the stock market and my e-mail via computer. I am writing this book on a computer and will send chapters for review electronically. I paid our real estate tax bill via the Internet. And so on. New smart phones can perform many of these tasks—as well as send videos and photos.

The Internet has altered virtually every major form of media, from books to newspapers to music to movies—all in a breathtakingly short period of time.

Take as one marker, one starting point of the digital and information revolution, the first fully assembled desktop computer, the Apple I, which was sold in 1977. In the past 30 years, computers have gotten cheaper, faster, with much better memories. The number of personal computers in the world reached one billion in 2008. Robert Darnton, now Harvard's librarian, summarizes the recent breathtaking pace of the ongoing information revolution: from early writing to moveable type more than 5,000 years, from moveable type to Internet 524 years, from Internet to search engines 19 years, from search engines to Google's algorithmic relevance ranking, 7 years. Darnton comments, "Each change in the technology has transformed the information landscape, and the speed-up has continued at such a rate as to seem both unstoppable and incomprehensible."[1] Darnton further simply states "information is exploding... furiously around us and information technology is changing at...a bewildering speed...."[2] In 1965 Gordon Moore,[3] the cofounder of Intel, famously predicted that the number of transistors placed on an integrated circuit would increase exponentially, doubling every two years; this accelerating trend correctly describes the improvement of processing speed and memory capacity of all digital electronic devices of the past 40 years, a storm of technological progress and change. Even if people don't possess a computer they cannot escape the omnipresence of computers in daily life, as retail stores manage inventory, banks track our accounts, and libraries catalogue.

Computers and the Internet, however, of themselves are neutral. What has changed, because of information technologies, is our relationship to information. Technology provides the means to create, store, and transfer massive amounts of data. The issue is *not* the data itself but how to manage so much information and tailor it to our needs.

How much information? Estimates on the rate that information is surging take one's breath away, because, even if inaccurate, they suggest the sheer size of the explosion of digital information. One IBM study of 2006 estimated that by 2010 the "amount of digital information in the world will double every 11 hours."[4] Information here refers to content in the form of digital documents, images, and various forms of multimedia. Companies that supply Internet access and fiber-optic networks, such as Verizon and Comcast, have to scramble to maintain a growing capacity to carry such heavy demands.

So rapid is the growth in the global stock of digital data that author Kevin Kelly writes, "the very vocabulary used to indicate quantities has had to expand to keep pace. A decade or two ago, professional computer users and managers worked in kilobytes and megabytes. Now my students use laptops with hundreds of gigabytes of storage, and network managers have to think in terms of the terabyte (1,000 gigabytes) and the petabyte (1,000 terabytes). Beyond those lie the exabyte, zettabyte and yottabyte, each a thousand times bigger than the last."[5]

This flood of information of all kinds is changing how we think and how we deal with information. Many people I have spoken to now agree that their attention span is shorter. Reading material online tends to promote a style of scooping and scanning the content. This is somewhat different from how we read a book. I turn to this issue in chapter 2.

2. Communication, Entertainment, and Stimulation

We live in a world of electronic screens. Glowing screens conveying information, entertainment, and infotainment snag our attention no matter where we are—at work in front of the computer, on our smart phones, in front of our high-definition television, in supermarkets, and in health clubs, where screens display advice or advertisements. The eye-catching quality of modern screens compel us to watch, and the small glowing screens of phones, large and small, communicate, or try to, with us—e-mails, text messages, instant messages, call identifications, advertisements. Advertisers bombard most of us continually in new ways designed to catch our attention, in our work and professional worlds, in supermarkets and health clubs. Easily used devices enable us to listen to sound—music, of course, and the human voice. But images seem to have a special power.

When did the on-screen revolution begin? To limit the discussion here to our capacity to capture, store, and transmit dazzling images, let's (somewhat arbitrarily) set the late 1940s and early 1950s as a starting point, when the first television sets were sold. These early sets had tiny screens, with flickering black and white images. If we leap ahead to the present and glance at a large-screen high-definition television set, the contrast is striking. The crisp images and intense colors compel our eyes to look, even if the story content on the screen is not as compelling.

The underlying issue is what I call the *attention economy.*[6] If we think of the world of information as an economy and information is so available, what is in short supply is human attention. What is in short supply becomes increasingly valuable. Our attention is under assault from many sources, and we can easily and often feel overwhelmed by the imbalance of data and information versus attention. In spite of efforts at multitasking, we can still only focus attentively on one thing at a time.

The information technologies that clamor for our attention use the same silicon chips as my computer and HDTV. The images that pour forth create an intensely stimulating environment, far more stimulating than at any time in history because the impressions and impact cascading down on us are so vivid, so rich, so intense, so unrelenting—unless we just unplug them. Because of the gradual introduction of recording technologies and image-reproducing technologies over the past two centuries, we live in an information-saturated world. Our experience is to live amidst intense images and sounds, or in the electronic and digital sense, streams of data or streams of zeros and ones. When I shop in a food store, monitors flicker with ads and cooking shows; at home and at work, computers perch on desktops; high-definition images gleam forth from televisions; my cell phone's tiny screen provides information about who's calling me; sound proceeds from hand-held devices or from the surround-sound in movie theaters and rock concerts. Many, many stimuli, visual and aural, clamor for our attention. Indeed, a modern person registers a hundred times more sensory impressions than an 18th-century citizen—whether it is through the window of a speeding train or the windshield of an auto or via television or a computer monitor.[7]

Does reality itself look this good? If we watch a football game on a high-definition TV, modern graphics and video techniques provide riveting visual stimulation, because of replays, close-ups, digital graphics, and many other features. I am amused as I observe the athletes on the field, be they football players or baseball players, looking up at the giant stadium screens to watch themselves, in replay, score the touchdown or hit the homerun. The visual technology trumps reality, almost, because it can freeze time, reverse time, and be replayed in vivid color.

Everywhere I look I see how we are wired electronically to the world and other people. When I take the train to work, I notice many commuters listen to iPods, gaze at iPads or some other electronic tablet, or text on cell phones. When I get to work, open up my computer, and first deal with my e-mail, I commonly have 50 messages or more. Occasionally I will go online and look at YouTube, a visually very stimulating website with regularly moving icons that keep changing as an indicator of their status as favorites for those currently or just recently checking on these brief videos. When I go home at night I see visual advertisements pasted on the sides of trains and buses. When I get home I confront a pile of mail, much of it junk mail consisting of fliers, catalogues, and other various forms of information from people trying to get me to buy something or attend some concert or travel some place or attend some conference. If I watch television in the evening, hundreds of channels are available through my cable carrier. And it has been said that the average person is exposed to perhaps three thousand advertisements daily, from the different technologies used by marketers.[8] If I go out to the movies, high-fidelity sound from large speakers engulfs me and powerful images wash over me.

Let's distinguish between stimulation and information. Stimulation is a form of sensory information. But not all information is equal. The home page on my computer (msn.com) offers me more than 50 info-bits, nuggets, and factoids under various categories: news, entertainment, sports, money, lifestyle, weather, market update, local events, celebs & gossip, "must see," and "more." The site assures me there are constant updates, most of which contain information as entertainment that I will remember for scarcely an hour. This differs from information as knowledge or wisdom that stays with us and can guide us.

But more in chapter 2 about these changes and how they alter our thinking.

3. Changes in Work, Work Changing Us

The information and digital revolutions have a profound impact on our jobs and how we do our work. Work is changing and changing us.

How many jobs have you had? The U.S. Department of Labor estimates that one in two American workers have been with their current employer less than five years, that one in four workers have been with the same employer less than one year, and that employees today by their 38th birthday will have had between 10 and 14 different employers.[9]

Many reasons exist for the churn in the world of work. Not least is the rise of industries and professions that generate information, and are based on new informational technologies that use or produce computers and images. Many jobs and professional fields simply didn't exist 10 or even 5 years ago, especially those in the information economy, such as e-business and the technologies that service computers, networks, and high-tech equipment.

As an exception to this flux in the work world, I am an anachronism of steadiness, having taught in the same college for a third of a century. While I have stability where I work, much of my work has changed and will continue to change. For example, when I began to teach, I used chalk and a blackboard and lectured from handwritten notes. Now before class I sit in front of a computer screen, on average, about three hours a day. I answer e-mail, consult my college's electronic bulletin board, toil at my academic writing, prepare PowerPoint slides for classes, and post assignments on

WebCt (a digital bulletin board where students can retrieve articles, find their grades, and make comments in electronic discussion groups). Oh, yes, and I still stand in front of students in a classroom. But what a classroom, equipped with devices that can display PowerPoint and DVDs, live television, images projected on a screen from an electronic textbook!

It makes sense that the information and digital revolutions have transformed institutions that deal with knowledge and information. Of course, some practices of colleges and universities will chug along the same track as 30 years ago, with other aspects bringing more change—such as distance learning, new uses of software in education, and new academic majors emerging all the time. Even aside from the impact of computers, the college where I work keeps vital and transforms itself with new professors, new procedures for admissions and submitting grades, new students, new courses, new deans, and so forth.

Change also alters other areas of the work world. My wife got directions to a new restaurant via her cell phone, which also provided global positioning system (GPS) guidance. When we arrived at the restaurant, our waitress punched our order into a hand-held computer, which sent the order directly to the kitchen and other stations so that our drinks and meal arrived quickly and efficiently. The bill went on one of our credit cards, and the credit card industry itself depends totally on computers and the quick transmission of data.

My children already have had more jobs than I have had and sit in front of computers even longer during the day than I do. My older son also carries his work with him, via laptop and cell phone, wherever he goes. Indeed, the new technologies make it hard *not* to work, whether it is on the weekend or vacation or late at night.

4. Changed Social Behaviors and Values

New technologies give birth to new behaviors and norms for proper behavior. Just as the computer changes how people work, another useful tool of the information revolution, the cell phone, has changed how people behave and relate. People not infrequently will take a cell phone call even in the middle of a conversation: "Excuse me, I have to take this call." Public announcements in both concert halls and movie theaters remind us to "Please turn off all cell phones and pagers." But I still haven't gotten used to sitting next to fellow passengers on the T (which is how Bostonians refer to the trolley and subway) or commuter train who may be having an intimate cell phone conversation. Yesterday the cell phone user was a young woman conversing with her physician about a highly personal issue. Obviously she took the call in a public place because that's where she was when he returned her call. My students who are often looking for internships or part-time jobs take callbacks whenever and wherever they can—an advantage of the portable phone. I don't allow cell phone calls in class nor do I allow texting in class because I don't want students to lose focus on the discussion or the lecture. But texting has become so addictive that some students leave class "for the restroom" to send texts.

There is no disputing the value of new instant communication devices, but it takes some time for society to smoothly integrate them and develop new rules for their use. A ringing cell phone should not interrupt a formal dinner party, unless that phone belongs to the secretary of state or a brain surgeon on call. But, a confession: when my wife and I are entertaining at home and our phone rings, I still get up from the table to check our caller ID. My wife does tell me, however, to just let it ring: "They'll leave a message—or send a text." Old reflexes die hard. And then there is a first-year student of mine who told me that she IMs or texts or uses her cell phone from 60 to 80 times a day; another student, a senior, claims he uses his cell phone 4,000 times a month. He was kind enough to print out his cell phone bill for me. In one month of 2010 he actually texted 4,280 times. Students text in class, in spite of rules against this distraction, and they text during graduation exercises.

Decades ago families had a home phone, a land line in the current parlance, usually one phone for the whole family. How did we get from that clunky black rotary-dial phone to sleek razor-thin smart phones? The different styles and expanded functions of these telephones mark the sharp upward curve of the telephone's evolution, which serves as one gauge of the extent and speed of changes in communication and our behavior. The standard AT&T phone remained unchanged for decades until AT&T introduced the stylish Princess phone in 1959, a compact phone convenient for use in the bedroom because it had a light-up dial. From the first telephone to the Princess phone was 85 years, 20 years from the Princess phone to the first brick-shaped mobile phone, 10 years from the early cell phones to the smart phones like the BlackBerry, 5 years from the BlackBerry to that marvel of multimedia functions, the iPhone. These recent smart phones offer many additional features—GPS capability, cameras, media players—which are quite different from the voice-to-voice function of the original phones. Today a customer needs to download an instruction manual to fully exploit the features and apps of these devices. Are they still phones? Of course, but they've changed and they allow us to do much more in our lives. Recently my sister-in-law e-mailed me from her smart phone a brief video of a dinner party she was hosting. Bored on the way to work? Play a game on your phone or watch a video or check your e-mail.

A schema is an individual's mental representation that holds his or her understanding of a particular event or situation. If we extend this concept to cell phones, for example, we are constantly learning new schemas each time we upgrade our phones. One observant parent noticed his four-year-old daughter using a pincer movement with her two fingers on the page of a book, trying to make the words bigger the way she saw her mommy doing on her smart phone. Right motion but wrong medium. Another child was observed holding her mother's makeup compact to her ear the way she saw her daddy use his clamshell cell phone.

Careful observation of children readily reveals their apprehension of the manual and cognitive skills needed to use the thousands of everyday items that are part of life—from tape dispensers to can openers. The lesson is how embedded we are with our technologies, tied in to the culture.

When technological innovations work as designed, they are a delight and astonishing—but they also have glitches and sometimes pernicious viruses. All of us have confronted technology gone awry, and usually it provokes enormous frustration. Not all devices are self-evident in how they operate or are installed or made to work properly. Surely some readers have approached the level of despair and hopelessness I faced when trying to install a wireless fidelity (wi-fi) router in my home.

We also need to remind ourselves how truly amazing these technologies are—how they enable us to do undreamed-of feats and expand our sense of what is possible.

5. Speed and Altered Time

A common response of friends and colleagues to my queries about how things are going is "Fine, but I'm so *busy*—I don't have enough time." Even my students seem to have the experience of not enough time. The information revolution fosters the speeding up of our sense of time—and the shrinking of space. Cell phones let people know where we are: "I'm just leaving so I should be there in four minutes."

Many professionals feel pressure to work all the time. Insecurity or competitiveness or ambition may drive them, but wireless phones and wireless Internet connections and lightweight laptops allow a blurring of time so there is no time on or time off; there is a shift in meaning about what vacation means—namely that the employees can vacate the office physically, but not mentally.[10] My college's e-mail system has a feature that allows one to check to see who else is online at any given time; it staggers me to see the number of colleagues and students who might be using e-mail or the Internet at all hours of the night. Of course, I too am guilty of that since I am doing the very same electronic blurring of work time and leisure time.

Has time changed? Is it running faster? Time has speeded up for me, and I conclude that there are four reasons for this. First, and least important, is my chronological age, since each additional year is a smaller part of my whole life and thus a year seems shorter. (When I was 2, a year was half a lifetime; at age 50, a year was 1/50th of my life.) Second, the information revolution surrounds us with speedy devices. I am trying to do more within a clock hour; and trying to do more, to multitask, is to make time seem like it is going faster. The Internet allows me to quickly find articles and information without going to the library; to order what I want online rather than driving to a store. Thirdly, my computer has speeded up enormously and gets me accustomed to having e-mail and other digital tasks happened at lightning speeds. My computer, which is four years old and cost about $1,000, runs at 2.20 GigaHertz, which is faster than a $10 million Cray Computer of 20 years ago. E-mail means that decision making and doing some business can be speeded up; there is no waiting two or three days for snail mail to deliver a letter. Google searches are accomplished in milliseconds. Lastly, we live in a media culture that makes time precious and scarce. I've grown impatient to the degree that waiting for an image to download or waiting online in a store or being caught in traffic makes me feel like I am wasting time. I have learned the cultural lesson that time is money: Don't *waste* time, but *spend* it wisely. I have a sense of time-theft if I am caught in a useless meeting or trapped in a situation other than one in which I could be more active and more productive. Time pressure is the press to get things done. As I look around my world the messages are clear: fast food, the Fast Lane or EZpass for the turnpike. I hear on talk radio announcers commonly saying, "We just have a few seconds left," or "Make your comment brief as we are running out of time." I was trained as a therapist to be careful about time and used the phrase "We are out of time" to end sessions with clients. The very modern sports of basketball and football are played by the clock so that in my very relaxation as I watch them I am reminded of the tyranny of time—"Only four seconds left in the half!" or "The referee is putting two seconds back on the clock."

Thus it is the daily contact with computers and electronic media that push time to run *subjectively* faster.

6. Family and Personal Relationships

It's not the information revolution itself that has brought changes to the family and how people relate. Biological and emotional needs of people are not going to change. What has changed is the quick and vivid transmission of information, via movies and television, about what people are doing in their intimate personal lives. The improved technology of birth control contributed to the so-called sexual revolution of the 1960s. The media, with vivid images and stories, contribute to changing values and acceptance of alternative lifestyles. People have more information about the relational and sexual experiments of couples, whether they are dropouts, celebrities, or the neighbors next door. One result: it is more common than not for young couples to live together before marriage—if they ever do get married.

Before turning to U.S. Census Bureau statistics to confirm the impact on the modern American family, let me illustrate some changes by briefly looking at the weddings of three generations of a family I spoke to while writing this book.

Start with the young couple whom I am calling Adam and Eve to protect their privacy. Adam's grandparents came of age in the 1930s, before the information revolution and the sexual revolution. These grandparents, from large families, were married by a priest in a Catholic church in the late 1930s. The few black-and-white photos show the couple dressed in traditional formal garb at a small reception of perhaps 30 or 40 people. Adam's parents, the second generation, straddled both revolutions (sexual and information), and got married in the early 1970s. Their wedding was less traditional, with a liturgical dance to a strumming guitar as the highlight of the ceremony. The bride's father, a traditional Catholic, was unsure if his daughter actually got married. The color photos reveal the bride's dress to be pink and more hippie in style than traditional.

The third generation, Adam and his bride Eve, came of age during the 1980s and 1990s when the information revolution and the Internet were permeating the culture. The role of the information revolution in the wedding of Adam and his bride? Very different from that of their parents and grandparents in every way. The couple's wedding in the early 2000s was outdoors in a distinctly nontraditional ceremony. The tools of the information age—cell phones, computers, video and digital cameras—shaped and encompassed this wedding. The young couple had viewed a variety of weddings on television and surfed the Internet for ideas. The register for wedding gifts was online, and they completed many of the arrangements by e-mail and cell phones. The video camera and still photos were digital, allowing for instant inspection and viewing. But more than these technological changes were the social and behavioral changes. The young man and his bride had lived together before they were married—not particularly new behavior, but more public and commonly practiced. The couple asked a relative, a noncleric, to preside at the wedding, enabled by Massachusetts law, which allows for a family member or friend to secure a one-day commission to perform a marriage ceremony. The best man was gay, and there was an openly gay couple at the reception who are planning their own wedding, as Massachusetts is one of the states that allows for the legal marriage of gay couples.

7. Contradictory Social Trends and Fewer Shared Experiences

The tools of information and communication make all of us more aware of the cultural discussions, trends, arguments, and innovations that go on around us. The better one is informed, the dizzier one feels. So much, so fast, so little time to register what is occurring and to respond. One can use a smart phone to find the name of that movie that's playing, as well as the schedule of screenings. Electronic readers enable us to download that new book in seconds. One can watch a movie on a smart phone.

Amidst this abundance of information is a fragmenting of experiences. People share fewer and fewer common experiences. In the 1960s and 1970s people tended to see the same summer blockbusters and get their news from the same television networks and same papers. A few big hits in music, movies, and books tended to dominate the mass culture in the United States. Movies and television shows of the time, comments author Chris Anderson, were "more popular . . . not because they were better but because we had fewer alternatives to compete for our screen attention."[11] Broadband, Mp3s, and online resources now push to the sideline traditional media and entertainment companies. Now teens and young adults swim in a very different media world, one that gives unfiltered and unlimited access to stimulating content of all kinds, from mainstream to the fringes of the culture. The Internet transforms so many markets, in part, because storage costs are so much less than for brick-and-mortar stores. One can find and rent movies that never reached the local cinemas. Sports equipment or specialized clothing can be found online. No store can do what Apple iTunes does: have available 2 million songs at almost zero cost of storage and instant access for a reasonable price. A Web retailer can make a profit by selling a few of many choices: Netflix rents about 95 percent of its more than 50,000 titles and Amazon appears to sell at least one of 98 percent of its top 100,000 books.[12]

The consequence is more and more choices, which is good, but there are more and more niche and specialized markets. Colleges offer much more diverse offerings rather than the traditional core courses on Western literature and Western history. It is good that students learn more about the rest of the world, but then they go about their separate ways and it is unpopular to require common experiences.[13] Three television networks used to offer limited programming, but the very limitation also meant that there were more common experiences. Knowledge has continued to become more specialized, which means that cultural literacy is quite different now than when people read the same books or listened to the same music.

With life more complex, people interact differently. Often people are removed from direct experience of each other, connecting via disembodied voices on the phone or brief text messages or e-mails that do not allow for emotional nuance. Although more connections are possible via electronic means, people seem to have fewer face-to-face encounters. Seeking to capture the American emphasis on individuality rather community, Robert Putnam used the phrase "bowling alone,"[14] even as other researchers seek a more nuanced view, as we shall see in chapter 7.

Sports have grown ever larger in American life as a new secular ritual for sharing a common experience.[15] One's favorite team provides an as-if community and forges ties of loyalty marked by the wearing of jerseys with a team logo or an athlete's number on the back. Partying and watching the Super Bowl—with its ads—each January provides a digital tribal experience for millions and grist for Monday's water-cooler conversation. A fan's team forges ties between people in ways that previous common cultural experiences used to provide—shared books, shared movies, shared TV programs. Of course, these are still present, but now there are many, many more options for individuals—blogs, favorite websites, hundreds of cable channels, the capacity to order up and watch thousands of past movies.

With fewer common connecting links across society, it is little wonder that American politics has grown more partisan and fractured rather than finding common ground. This is, of course, a complex issue, but there is a cultural argument going on in America. People feel the culture is in flux, and many are upset and angry. The argument in many ways is a culture war—whose America are we living in? In part, the technological changes have precipitated the fight, or at least have transformed the way in which political and cultural arguments get debated.

Confronting the Meaning and Consequences of these Changes

I have had one employer for the past 33 years. I have been married to the same woman for almost 35 years. By temperament and upbringing, I don't particularly *like* change, but in order to cope positively with change I decided to look it straight in the eye and try to figure it out. Unease and curiosity impel me to carefully scrutinize change and seek some ways to understand it in its many forms.

I notice change in my life because I have been alive for six decades, a span of time that exposed me to the big markers of modern technologies: the arrival of new phones, the first appearance of computers, the beginning of e-mail. These experiences give me a baseline, a way of measuring or comparing experiences and objects. When I encounter a new style or notice a new piece of technology, there is an initial sense of strangeness, of something alien to my usual sense of the familiar and safe. The perception of otherness, of differentness, probably is a survival reflex left over from evolution that alerts us to a possible threat versus the familiar. We, however, are an adaptive species and easily become habituated to new stimuli so that we scarcely notice them unless we intentionally attend to them. An example would be how quickly my sons and my college students simply and casually adjust to the new tools of the information revolution.

Today's young people, like my granddaughter and students, are just beginning to put their toes in the moving river of culture and so accept their experiences as a given, the way things are. This is life and reality for them. As they build memories of the familiar, they begin to establish a baseline of experiences. My three-and-a-half year old granddaughter, with her play cell phone and a toy computer, will grow up accustomed to these extraordinary tools. To be sure, there are differences in the generations and differences in the technologies that each generation encounters.

How Are the Changes of the Information Revolution Changing us?

Behaviors, Brains, and Ways of Thinking

Change occurs not only in the externals, the tangible stuff of life, but also in our mental attitudes and categories and ways of thinking—and our brains. Of course, modern people think differently in a variety of ways from how our ancestors did, but many of these changes in our thinking are occurring right now. For example, our sense of the real is altered. Are we watching a replay on television or is this live? Is it actual or is it digital and previously recorded? Baseball players pause in their home run trot around the bases to glance up at the stadium Jumbotron screen to watch themselves hit the ball seconds previously—time has been reversed. At my son's wedding, digital images of the ceremony and video clips were ready instantly, frozen forever for future viewing. Surely this must alter our attention span, and it's no surprise that some complain they can spare little time for a 500-page novel as they rush to the health club, answer calls on their cell phones, and respond to dozens of e-mail messages.

The availability of experiences and the unprecedented pace of life pace have a range of consequences. Coauthors John Robinson and Geoffrey Godbey comment, "Americans can experience more than any generation in history. They can do more, see more, travel more. The repertoire of experiences for people of means can be enhanced almost indefinitely."[16] This expansion of experiences works well for many people, but many also feel that the technological tools of modern life are exacting a cost in the form of diminished quality of life, with diminished privacy and greater stress.

In a later chapter I shall review the research on how our very brains are changing.

Machines that make use of new computer chips and digital displays have led Boston commuters to change how they pay to ride the Metro trains or the T. New machines allowed for the introduction of paper tickets to replace cash fares. Commuters adjusted, slightly altering their patterns of how they purchased tickets or Charlie Cards prior to boarding the trolley cars. They also had to slightly alter their thought frames and assumptions: they did not need a pocket full of change but could purchase a ticket from a vending machine with a credit card. Or they could plan ahead and have their employers arrange a monthly pass. A daily commute became slightly more complex, but more convenient—another small change that people adapted to.

Another illustration of new behavior and consequent new ways of thinking is from a class of my freshman: of 23 students in the class, 22 did not have a watch. How did they tell time? They used their cell phones, of course. This illustrates not only a new use for the evolving technology of the phone but also a new mental schema (a way of mentally interpreting and organizing experience), so that when there is the need to know the time, one looks at one's phone rather than one's wrist. (A curious consequence of widespread use of cell phones for time-telling is a decline of 17 percent in the sale of watches over the past five years.)[17] A further example of a new cognitive frame is the student who left a voice-mail message on my office phone, without leaving a callback number; she assumed that the number she had reached was a cell phone that would capture her number and enable me to respond. Alas, my office land line had no such sophisticated feature and her call went unreturned.

Paradigm Shift

The digital world and cyberspace information have changed how people think, by providing—imposing—new categories and assumptions. Changing how a person thinks changes the person. All of us integrate our experiences into various cognitive and logical structures, ways of interpreting our world and organizing responses. To watch my old television set I would walk into the room and push a button on top of the set; with my new television, I pick up the remote control and press the on button. It's only when I unthinkingly walk into the room and try to press the top of our new HD set that I realize the different technology has compelled me in a small way to change my behavior, which in turn required a small mental adjustment: I have acquired a new mental structure or pattern for the changed situation.

Psychologists refer to these mental structures as schemas. A few examples will serve to illustrate schemas. A football player has one set of thoughts and responses when he plays defense. If the coach switches him to offense, he needs to develop different mental reactions and ways of moving his body. Driving a car with automatic transmission requires slight shifts in one's schemas from those used in driving a standard transmission. Recently my health club changed how they located the exercise machines. This required me to make a slight change in my mental map and schema, as I altered my exercise routine based on progressively moving along a particular row of machines. We have countless of these small mental schemas that only alert us to their presence when they are outmoded or need to be changed because, perhaps, of a new computer or new software.

In the past when I wished a phone number or address for a store, I picked up the phone book and flipped through the pages. At times a different schema might prompt dialing 411 for information. Now, when I require the phone number and address for the nearest Radio Shack, I Google

"Radio Shack" to get not only directions, but store hours, distance to the store, and so forth—much more information than I need, but the advantage is great.

Larger and more complex schemas, called paradigms, are changing. The term *paradigm* comes from Thomas Kuhn[18] to describe a cognitive change in how we understand a larger issue—even a worldview. Kuhn used the phrase "paradigm shift" to refer to a new explanation for anomalies outside a reigning scientific theory. Such a shift indicated a substantial change in assumptions and understanding. When, for example, Copernicus determined the earth revolves around the sun rather than the sun around the earth, it was a paradigm shift. I believe that world of the Ethernet or Cyberspace, unexpected and unforeseen prior to the advent of the computer, sets up a new paradigm of information and communication. In spite of the overstatements of some enthusiasts who ascribe transformative power to the Internet, claiming that it will unleash populist forces, the Internet has changed how the modern world does its business, in large and small ways, from thinking how to arrange an airline flight to getting instant information about the stock market.

Why Bother Thinking about Change?

What would happen if we were to ignore the changes around us? If we expect to live in our culture, we must come to terms with rapid change. To ignore or avoid change is to be left behind. Let me illustrate by recounting the story of a man who amazingly did not change for *19 years* and then had to face a world vastly different. A 65-year-old Polish railwayman regained consciousness after being in a coma for 19 years. In 1988, Jan Grzebski was in an accident and doctors estimated he would only live two or three years. At that time Poland was still communist. For 19 years Gertruda Grzebska, his wife, "did the job of an experienced intensive care team, changing her comatose husband's position every hour to prevent bed-sore infections," said Dr. Boguslaw Poniatowski.[19] To the amazement of his family, his doctors and the world, Grzebska regained consciousness. The wheelchair-bound Grzebska said: "When I went into a coma there was only tea and vinegar in the shops, meat was rationed and huge petrol queues were everywhere.... Now I see people on the streets with cell phones and there are so many goods in the shops it makes my head spin."[20] This modern-day Rip Van Winkle awoke to find a world dominated by computers and speedy information devices; to his confusion and amazement, a digital revolution had arrived. His disorienting experience illustrates the theme of this book: that things change constantly and also result in people being transformed—or not, in his case. If each of us were somehow not paying attention, adjusting, making adaptations, the sense of change would be overwhelming and bewildering. Our heads would indeed spin.

Oh, the Complexity!

The dynamic complexity of modern culture offers fertile soil for innovation and experimentation, the new and different. We are all faced with the necessity of confronting changes in social behaviors, etiquette, new ways of doing things. The petty and profound get mashed up. What is the proper way of answering the phone when you have caller identification? What is appropriate cell phone etiquette in public? How do e-mail and Tweeting alter people's relationships? How do computer use, texting, and e-mail diminish the quality of handwriting? (We know for sure that it contributes to the decline in letter writing and the consequent decrease in first-class mail volume by hundreds of millions of pieces.[21]) Do people write thank-you notes or reply to the RSVP request in the manner that an earlier generation was taught? Are new forms of civility, social rituals, and manners emerging? What are the resources we need to organize the kaleidoscope of events and experiences in our lives? How should we think about, and differentiate among, these changes and fads? What resources do we reach for to swim in this torrent of change and difference?

There are, of course, those who are on the sidelines in this rushing, changing culture, who have dropped out of the race or are determined to resist the changes they see. They have economic or religious or philosophical or political reasons. The poor who are scrambling to survive do not have the luxury of staying current with pricey technological gadgets. Some individuals resist change by clinging to what they hope is steady and continuous. I interviewed a man who does not possess or use a cell phone or computer, and he does not have credit cards. He acknowledges that this makes life and travel in the contemporary world sometimes awkward. Some friends and colleagues and neighbors are puzzled and troubled as they grapple with new experiences, marvelous innovations, and bewildering new pressures.

How do we manage these changes? How are these changes changing us? This book is an exploratory trek into the heart of current American life, to explore the very stuff of life evolving and changing as we experience it. We have grown accustomed to some of these resources, but they do increasingly alter the cultural landscape that existed just one generation ago.

Many of the changes I face are positive and make my life more convenient and pleasant; some cause stress and anxiety because they demand that I adjust and adapt. Before turning to the management of changes and an understanding of how they are changing us, I wish to look more deeply into areas where change is especially present.

2

So Much Information Is Changing How We Think

Going, Going, Almost Gone

I used to enjoy the ritual, now rapidly growing extinct, of spreading out three morning papers on our kitchen table and reading them as I ate breakfast and got ready to leave for work. Replacing that ritual are many channels of information pouring forth torrents and torrents of data, images, and up-to-the-minute information. In daily life this reality looks something like this: glancing at a television in the kitchen each morning to break the silence and catch the weather or highlights of last night's Red Sox game; listening on the commute to an iPod for music or for the latest news or an uploaded book or recorded program; checking the smart phone for text or messages; arriving at work, and turning on the computer for the Internet to check on e-mail or the current Dow Jones numbers or surf for some amusing video clip on YouTube or a wall posting on Facebook. If I need to look up some fact, perhaps I'll make a quick stop at Wikipedia, or check tennis scores at the Wimbledon site, or a friend's blog.

The devices of the information revolution surround us: laptop and desktop computers, compact video-recording cameras, multiuse smart phones, and ubiquitous flat-panel monitors at work and at home and in stores. Of all these tools to create, capture, and reproduce information and images, the most compelling of all, however, is the computer connected to the Internet. All I need to do now is move my fingers to the on switch of my home computer—or computer tablet or smart phone—to access quickly the World Wide Web and unleash this gush of information.

This fountain of information is more than the news about politics or sports or current events; it is the availability of facts and images of almost anything I could want to know. With a few keyboard strokes I can access hundreds of journals and tens of thousands of journal articles from the past 80 years or so. Blogs? Tens of thousands. Find art? I can visit many of the museums of the world and see crisp, color images of their collections on my flat-screen monitor. Some museum websites allow for a 360-degree view of a particular vase or statue, just as one can visit an apartment for rent or a condo for sale through real estate sites that provide a slow gaze around each room. I routinely use MapQuest for directions, or to search for a product or store location. I use Google to locate the address of the nearest retail outlet and to find its store hours. Available online are sites that offer medical advice, help on getting rid of things, or answers to queries on how things work. Want to play poker? There are sites. Want puzzles or jokes? Choose your sites. Need information about Chinese etiquette? Want to know how to mix a Singapore Sling? You can find it on the Web. On the radio this morning I heard a song sung by Theresa Brewer, a pleasant but minor singer from the 1950s. I had been a teenage fan then and so I Googled her and, of course, I quickly found biographical information, pictures, and lyrics to her songs. This book could not have been written without access to the resources of the Internet.

Information technologies touch virtually every aspect of our lives—"how we communicate, entertain ourselves, seek information, make a living, connect with family and friends and pursue our life goals. . . . Look at the average 14-year-old today and see how technology surrounds every aspect of his or her life every day."[1]

Information is no longer scarce. New ways to access information are replacing the old. The challenge is how to filter and manage the force of so much digital and electronic information.

Information and Our Changed Relationship to It

The electronic devices and the Internet enable us to have so much information at our fingertips that it changes our very relationship to information.

The current revolution in Information is comparable to how Johannes Gutenberg's printing press around 1440 altered people's relationship to knowledge. But Gutenberg's impact took hundreds of years to be integrated and spread among people. Almost two thousand years prior to Gutenberg's and earlier Chinese presses, an earlier knowledge revolution was in operation. I am referring to the introduction of writing, which transformed communication and reliance on human memory. At that time Socrates, through Plato's words (*Phaedrus* 275a–b), warned that writing would change people's relationship to knowledge, and he was right. Robert Darnton, the Harvard University Librarian, raises the question of how we can orient ourselves in the current culture when "information is exploding so furiously around us and information technology is changing at such a bewildering speed . . . the pace of change seems breathtaking . . . Each change in the technology has transformed the information landscape, and the speed-up has continued at such a rate as to seem both unstoppable and incomprehensible."[2]

Our current revolution, based on digital technologies, has spread around the world in a mere matter of decades, a blink of the eye in the history of humanity.

The Size and Contents of the Digital World and Internet

How Much Information?

In the next several pages I want to suggest how unimaginably big the Internet and digital universe is so that we can better get a sense of how this information, readily available at our fingertips, is changing us. Many of the online sites no longer bother counting blogs or demonstrating the legitimacy of the Web by citing numbers, but since blogs and some of the other features of the Internet are still in the process of entering mainstream practice and consciousness, let's proceed.[3]

First of all, no one knows how large this deluge of digital data is, because the Internet was designed initially to resist failure or attack and thus there is no single control point. The Internet networking technology continues to grow anarchically without central planning. A Netcraft Web Server Survey reported more than 108 million websites existed in 2007, and this had doubled to 234 million by December 2009,[4] with some of the increase from the blogging service of a Chinese social network, Ozone. But the number of websites fluctuates wildly from month to month, and one runs into a problem of what exactly constitutes a website. Is a person's individual Facebook page its own website? How about blogs? Estimates of the number of pages on the World Wide Web range from more than 29.7 billion pages to more than 600 billion.[5] Eric Schmidt, the CEO of Google, the world's largest index of the Internet, estimated the size at about 5 million terabytes of data. That's over 5 billion gigabytes of data, or 5 trillion megabytes. Schmidt further noted that in its first seven years of operations, Google had "indexed roughly 200 terabytes of that, or .004% of the total size."[6] In 2010 the website Pingdom reported that there were 1.4 billion e-mail users worldwide, with 247 billion e-mails sent per day, which is far more than a whole year's worth of letter mail in the United States. There were more than 81 million .com domain names at the end of 2009 and 1 billion videos that YouTube serves in one day. Four million images are uploaded everyday to Flicker.com, enough to fill a 375,000 page album.[7]

Distinguishing knowledge from data on the Internet is enlightening, as the bulk of new data comes from the transmission and storage of images, streaming video, and text messages. Measuring the increase of knowledge by counting objects like books or journals or new patents gives varying estimates. Thus the number of new biology journals in print and in digital versions seems to be doubling every five years, and medical knowledge seems to be doubling every three and a half years—or is it every 32 hours, as one blog claimed in 2007?[8] Currently the world's store of digital content is the equivalent of one full top-of-the-line iPod for every two people on the planet. If we were to print and bind into books the world's current digital content, amounting to billions of gigabytes, the stack would stretch from Earth to Pluto 10 times.[9]

Between 2006 and 2010 the information—mostly data rather than knowledge—added annually to the digital universe increased by five or six times. If one considers that there are roughly one billion devices in the world that capture, store, and transmit images (cameras, medical scanners, security cameras, camera phones), the increase in data that conveys images substantially expands any estimate of the size of the digital universe.

Thus, the amount of digital information may be seven or eight million *times* the information in all the books ever written.

Those who attempt to estimate the size of the Web's content seem to relish these astronomical numbers of Internet texts, scanned-in documents, uploaded images, and videos. A digital expert, Doug Webster of Cisco Systems, says, "When we started releasing data publically, we measured it in petabytes of traffic . . . Then a couple of years ago we had to start measuring them in zettabytes, and now we're measuring them in yottabytes. One petabyte is equivalent to one million gigabytes. A zettabyte is a million petabytes, and a yottabyte is a million petabytes."[10] A few giant portals like Facebook, Google, and YouTube now generate 30 percent of Internet traffic, and Webster estimates that video will account for 90 percent of all Internet traffic within three years. How can we comprehend the sheer amount of data that is online?

About 20 years ago a writer claimed that a weekday copy of the *New York Times* contained more information than the average person in 17th-century England was likely to come across in a lifetime.[11] No evidence supported this claim, but even if it were true, the same claim is more doubtful now because of the shrunken state of the declining newspaper industry and the shrinkage of the *New York Times* itself and its advertising pages. As for that 17th-century Englishman, however, he did not know about, or need to respond to, famine in Africa, for instance, or a death 20 miles away. He did not possess multiple electronic sources of information telling him what was going on in the rest of the world; in his century, less was happening and news of it did not reach beyond those in the immediate neighborhood. In contrast, we have available more information than anyone can make sense of about bombings, hurricanes, and plane crashes around the world.

How Fast Is the Internet Growing?

Researchers seek to capture the surging growth of the Internet in numbers. They believe that the network grows in exponential form, similar to Intel founder Gordon Moore's prediction that the number of transistors on a central processing unit would double every 18 months. Moore's prediction has held up since 1965. The researchers, using data collected at

6-month intervals between December 2001 and December 2006, estimate that the Internet doubles every 5.32 years.[12]

How fast is that mountain of digital data growing? Some estimate that the growth rate of the world's knowledge is a doubling every four years...and others say it is doubling every two years.[13]

Adding to the surge in growth from smart phones and streaming videos is another engine of swelling data: the Google Books Library Project. This project involves scanning and digitizing millions of books in the world's premier libraries to create a vast searchable database of human learning. Libraries at Harvard, Oxford, Cornell, Princeton, Michigan, the Bavarian State Library, the Ghent Library, and the National Library of Catalonia are participating. At this stage Google workers are only scanning public domain material, that is, material whose copyright has expired. As the project progresses the range of material may increase to include journals, dissertations, government publications, and foreign language material.[14]

To use such a vast storehouse obviously requires a computer and access to the database. There are approximately 543 million books in the research libraries of the United States. Google will be able to digitize about 15 million books worldwide at current rates; but even if Google digitized 90 percent of the books in the United States alone, millions of books and documents will remain nondigitized. In our digital age, to be in a nonelectronic state means being invisible to computers. Thus, the nonscanned material will be lost to current and future digital researchers. Indeed, when the world entered the digital age, the vast majority of historical records did not make the crossover. How much casual or marginal literature, and how many scientific journals, films, court and legal records, corporate documents, letters, and diaries, all accumulated over centuries, will slowly vanish because they cannot be searched and archived with newer digital material? What may not be important to today's scholars might be crucial for some future scholars.[15]

Another example of areas of growth in the Internet is C-Span's decision to put online and allow access to more than 160,000 hours of C-Span footage. C-Span began in 1979 as a creation of the cable industry to record every Congressional session, every White House press briefing, and other acts of official Washington. This volume of videos is so vast that finding valuable references is like looking for a needle.[16]

Who Is Using the Internet?

Another measure of the size of the Internet is to estimate how many people use it. Several companies, including the Central Intelligence Agency, estimate that over a billion people used the Internet in 2008, with probably 500 million using it at least once a week. The Internet has become the 21st-century equivalent of the 19th-century railroad or the mid-20th-century's interstate highway.

Who in the United States is interacting with the Internet? In a national survey late in 2009, the Pew Research Center's Internet and American Life Project[17] finds: 74 percent of American adults (ages 18 and older, more than 200 million people) use the Internet, 60 percent of American adults use broadband connections at home, and 55 percent of American adults connect to the Internet wirelessly, either through a Wi-Fi or WiMax connection via their laptops or through a handheld device like a smart phone. The U.S. Census Bureau[18] offers similar figures, but broken down in a way to show that younger people and people with better education are far more likely to be connected to the Internet. The growth of the digital universe is uneven, however, with most of the activity in the Asian rim, North America, and Western Europe.[19]

Internet users can be divided roughly into *digital natives* and *digital immigrants. Digital natives*[20] were born after 1985, have grown up with digital technologies, and have familiarity and ease with these technologies, whether social networks, websites, or various applications of smart phones. They live much of their daily lives online, moving easily between online and offline. On the other hand, *digital immigrants* were born before the revolution and more or less have adapted to the changes brought along by computers and the Internet. Digital natives differ from digital immigrants in how they *feel* about technology, that is, technology is simply "'stuff' they use everyday. Immigrants are constantly amazed by technology."[21] Digital immigrants tend to encounter more frustration and stress when trying to download some materials for Skype or adopt new software. They use technology different from natives; immigrants use it to do the things they do more effectively—using websites to seek information, and so forth. Digital natives ride, and thrive, on a moving conveyor belt, whereas digital immigrants often feel they are trying to keep their balance on this fast-moving conveyor belt. Digital natives use technology more broadly, to be entertained and stay socially connected.

Most importantly, digital natives demonstrate how we have changed and what we are becoming. And the immigrants are looking more and more like the natives.

So Much Information Is Changing Us

Attention Spans are Shorter

First, my own experience. My attention span is impaired, growing shorter and more distractible. My electronic environment of the Internet (and television as well) distracts me and *trains* me to be more distractible. The Net, with its increasing screen clutter, such as pop-ups and embedded ads, seems to foster reading shorter passages, hopping from one site to another. My ability to sustain focus for longer and denser prose has eroded.

Nicholas Carr[22] describes how he is changing because of his use of the Internet and digital resources. He used to be able to spend hours reading long stretches of prose, but now his concentration drifts after two or three pages. He fidgets and looks for something else to do. He attributes this struggle to stay focused from the 10 years he has spent online, surfing the Web. He appreciates the advantage of quick searches on Google for quotes that would have required days in the library searching the stacks. He enjoys moving from one hyperlink to the next. Yes, there are many advantages of such immediate access to an incredibly rich warehouse of information. "What the Net seems to be doing," he says, "is chipping away my capacity for concentration and contemplation. My mind now expects to take in information the way the Net distributes it: in a swiftly moving stream of particles. Once I was a scuba diver in the sea of words. Now I zip along the surface like a guy on a Jet Ski."[23]

The information glut (thousands of data bases, abundance of scholarly journals, new sources of stimulation that I consider in chapter 3) turn our culture into an attention economy.[24] That is, what is scarce becomes valuable. With new technology offering ever more seductive and distracting advertisements, with convenient technologies sending us e-mails and text messages, with high-quality monitors in many locations, all these make demands on our attention. As a result we face the task of how to allocate a precious resource, our attention. Thus we have shorter and shorter attention spans as we try to allocate resources to all the stimuli that are coming at us.

If one commodity becomes abundant (information) and another scarce (attention, ability to focus), there is an imbalance. But time to deal with the flood of information is another matter. That requires attention. And attention is in short supply. The rich can't buy more. The combined shortage of time and attention is unique to today's information glut. Consumers are willing to pay handsomely to save time, and marketers are just as eager to pay to get the consumer's attention. This results in a decline in the quantity and quality of serious thinking. Author Ed Shane comments, "Useful knowledge however is in short supply, especially compared to information devoted to entertainment and commercial interests."[25]

The words *new brevity* seem to describe this world of expanding digital and electronic experiences. Tyler Cowen, a cultural commentator, describes it well: "The relative decline of the book is part of a broader shift toward short and to the point. Small cultural bits—written words, music, videos—have never been easier to record, store, organize, and search, and thus they are a growing part of our enjoyment and education."[26] Cowen further suggests that in the current cultural transformation, Google and websites are more likely to provide digital natives a formative cultural experience than traditional novels, so that books and traditional paper sources of information become less central to our cultural life.

Changes in Reading Patterns

Reading books in America is declining among all ages, the National Endowment for the Arts reported a few years ago.[27] A national survey revealed a dramatic decline in reading literature, with the steepest rate of decline, 28 percent among the youngest groups, those most likely to be using digital technologies. The study documents an overall decline from 1982 to 2002, with a sharper rate of decline in the last decade. Americans are reading less and less traditional paper sources of information, such as newspapers, magazines, and books. Newspaper circulation is on a negative downward slope. Last year circulation dropped 9 percent and this drop followed years of declining circulation.[28] The once robust *New York Times,*

Washington Post, Time Magazine, and *Newsweek* have grown emaciated. Traditional books seem to be less central to our culture as electronic readers like the Kindle and iPad, as well as computer screens, move to center stage.

How we read print on a screen is different from how we read print in a book. Eye-tracking visualizations[29] show that users often read digital screens or Web pages in an F-shaped pattern: a quick horizontal swipe across the upper part of content, then down the page a bit and again another horizontal swipe that usually covers a shorter area. Then the reader—often called a *user* in this context—scans the content's left side in a vertical movement. The implications of the F Pattern for Web design are clear and show the importance of following the guidelines for writing for the Web: we are told that users or readers won't read such text thoroughly in a word-by-word manner, that exhaustive reading is rare, that yes, some people will read more, but most won't, that it's better to have bullet points with information that users will notice when scanning down the left side of the content in the final stem of their F-behavior.[30]

Another distinctive change in how we read words online can be seen in hypertext. Traditional books and newspapers arrange words in linear text, which we read back and forth, from left to right, making our way line by line sequentially down the two-dimensional page. Online hypertext, by contrast, uses embedded links that take the reader to other places in the text body or to totally other texts. It is as if hypertext has virtually infinite dimensions that one can penetrate to multiple levels and innumerable sites. Hypertext alters the experience of reading by enabling the reader to follow a trail of information, jumping from one section to another, flexibly following a path more of association than of the linear logic of traditional texts.[31]

Researchers from the University College London carefully monitored how digital visitors to British libraries used electronic books and scholarly journals. They found that visitors manifested "a form of skimming activity," skipping from one source to another and rarely returning to any source they had already visited. The typical visitor read only one or two pages of an article or book before bouncing to another site. Many of these visitors were in their school years and therefore likely to access and interact with digital resources for years. These digital natives' approach appears to differ greatly from the ways that digital immigrant researchers and scholars work.[32] Carr claims, "It is clear that users are not reading online in the traditional sense; indeed there are signs that new forms of 'reading' are emerging as users 'power browse' horizontally through titles, contents pages and abstracts going for quick wins."[33] He speculates that with the availability of information on the Internet and even on smart phones, we may be reading more than before, but it seems to be a different kind of reading.

A survey by The Pew Internet & American Life Project (2010) seems to confirm Carr's view. In the Pew Report many information workers (researchers, teachers, etc.) say they feel stress at the volume of information available, and further they say their response is to develop a style of skimming over the surface of the material they are scrutinizing.[34]

If we read differently, our brains are operating in a different way. "We are not only *what* we read," says Maryanne Wolf, a developmental psychologist at Tufts University and the author of *Proust and the Squid: The Story and Science of the Reading Brain.* "We are *how* we read."[35] Wolf worries that the style of reading promoted by the Net, which puts efficiency and immediacy above all else, may be weakening our capacity for the kind of deep reading that emerged when an earlier technology, the printing press, made long and complex works of prose commonplace. When we read online, she says, we tend to become "mere decoders of information." Our ability to interpret text, to make the rich mental connections that form when we read deeply and without distraction, remains largely disengaged.

Reading, explains Wolf, is not an instinctive skill for humans. It is not genetic in the way speech is. The human brain has to learn how to translate the symbolic characters we see into the language we understand. And the media or other technologies we use in learning and practicing the craft of reading play an important part in shaping the neural circuits inside our brains.[36] The brains of those who read ideograms, such as the Chinese, have different mental circuits for reading than those of us who read and write a language with a phonetic alphabet.[37] Wolfe affirms that these variations extend across many regions of the brain, including those that govern the essential cognitive functions of memory and the interpretation of visual and auditory stimuli. Accordingly, she says we can expect the circuits developed in our use of the Internet are different from those formed in reading of books and other printed works. So while technology enables us to survey vast amounts of information and even do many things at once, this divided attention seems to impair our ability to commit to deeper engagement with ideas and sustained trains of thought. Cowen contrasts the intellectual and emotional satisfactions that come from engaging in the range of emotions of a Mozart opera, for instance, over several hours, with the assemblage of various jokes from YouTube, a song from iTunes, and a brief glimpse from a movie online.[38] In brief, many of us are indeed changing how we read, at least on the computer screen. It may make us more literate in a different kind of way. So the Internet may not be making us dumber, but just different, changed.

The Bottom Line: Our Brains are Changing

The different kind of reading on monitor screens points to changes in thinking styles.

The experience of searching through massive amounts of data and retrieving what we are seeking implies a more associative than linear logic. Those who grow up using digital technologies, the digital natives, have brain circuitry slightly different than those reading mostly traditional print. The heavy users of information technology become accustomed to short-term multitasking and parallel processing. Their access to visual and auditory stimulation has programmed their brains to crave instant gratification. Gary Small and Gigi Vorgan[39] confirm that the brains of the digital generation are different as they easily adapt to a technology-driven information culture that is overtaking yesterday's low-tech channels of information. The bombardment of digital stimuli on young minds teaches them to respond faster, which prompts the development of mental circuitry slightly different from older minds accustomed to a different kind of reading and attention.

Small and his colleagues used functional magnetic resonance imaging on volunteers aged 55–76 to explore the possible influence of using the Internet on the brain. They compared functional MRI scans on those who were engaged in Internet search tasks with those who were reading text on a computer screen formatted to simulate a printed book. The conclusions from comparing the brain scans of the two groups were that Internet searching did indeed alter the brain's responsiveness in neural circuits controlling decision making and complex reasoning.[40]

Interacting with Internet content is a very different experience from the sustained concentration required for traditional cultural forms, such as a four-hour opera or a long novel. Many more of us now are visiting the Internet frequently, and thus small cultural bits, be they words or music or video, are increasingly more central to our daily thought and emotional life. Tyler Cowen calls this our "attention-deficit culture" and refers to the "new brevity."[41] Thus the digital natives—and many of the rest of us digital immigrants—find that there is some rewiring in our brains so that we are restless and less attentive unless the stimuli are especially compelling, and this shorter attention span renders even conventional television too sluggish and boring. One-third of young people use other technologies, usually phones or iPods, while watching TV. Increasingly younger students multitask almost constantly, downloading music to their iPods and instant messaging their friends while doing homework. Young brains are much more sensitive to environmental input than are more mature brains. Younger minds are simultaneously more vulnerable to brain-influencing new technology but also more awash in it.

Researchers are learning just how adaptive the human brain is and how the brain adjusts to the demands of the environments. I mentioned above that those who write Chinese ideograms have differences in their brains from those who write alphabetic systems. In addition, Chinese (Mandarin) speakers and English speakers have slight but detectable differences in their brains. Using functional magnetic resonance imaging (MRI), neuroscientists have tracked how these two language groups use their brains differently.[42]

Perhaps the best illustration of the brain's plasticity is to compare the brains of musicians and nonmusicians.[43] The sustained practice and repeated motions of a violinist's arm and fingers result in detectable changes in the brain compared to those who are not musicians.

It is still too early to detect just how much the cultural shift from books to computers is affecting our brains. Perhaps the mix of different kinds of reading (the printed TV guide, the contents of the cereal box as well as what's on the computer screen) will keep our brains more or less unchanged.

The New Cultural Literacy

Will all this information and the Internet make us stupid? Nicholas Carr[44] asked this question, and the bottom line is that we are *different*. There is, willy-nilly, a new standard of cultural literacy emerging that proclaims that remembering information isn't as important as knowing how to access it and manage it. Tyler Cowen summarizes forcefully what cultural literacy has become: not familiarity with the plays of Shakespeare or knowing how to parse a Rubens painting, but rather knowing how to operate an iPhone and Web-related technologies. It is not, he says, "so much about having information as it is about knowing how to get it."[45]

Three variables come together: the shortening of attention span, insufficient training in sustained mental effort, and the expansion of new ways to obtain facts and information. The consequence? A rising tide of nonessential data, factoids, infotainment, the merely trivial and silly, and lastly misinformation and myth. My home page when I go online is msn.com, which offers the weather, breaking news, and many trivial items designed to amuse and divert. For instance, on one recent morning one could find out "Why this glacier appears to bleed," or "Dog wins ugly award," or "Find the perfect prom dress." The powerful technologies that enable users to upload video clips to YouTube, to be shared by tens of thousands, shelters many silly videos of men getting hit in the groin or pets doing tricks.

A recent book giving advice on Web design is, perhaps, overly frank in its title: *Don't Make Me Think: A Common Sense Approach to Web Usability*.[46] The author gives some common sense suggestions, but he puts in words what I have said above about the style of reading on the Web. On pages 22–23 he says, "We don't read pages. We scan them," and "We're usually in a hurry," and "We know we don't *need* to read everything." Lower intellectual expectations give confirmation to Susan Jacoby's concerns about debased standards and the least common denominator in much public discourse.[47]

The scooping, scanning style of gathering information online perhaps is necessary when faced with the huge bounty of data and facts online. But there is a transition to a new kind of literacy, from the expert who knows an organized body of specialized knowledge to the swift computer user who knows how to find things. The culturally literate digital native seems to have prowess in navigating to where desired information is. In everyday life this means knowing how to rapidly find websites that furnish directions, trivia factoids, quick online medical advice, or remedies for a new virus attack.

Changes in Education and Intellectual Life

Educators and cultural commentators are compelled to rethink the new information technologies and the very notion of information. Darnton suggests information should not be "understood as if it took the form of hard facts or nuggets of reality ready to be quarried . . . but rather as messages that are constantly being reshaped in the process of transmission. Instead of firmly fixed documents we must deal with multiple mutable texts."[48]

Education, the business of knowledge, tries to deal with the information revolution in a variety of ways. For example, I make use of PowerPoint—that omnipresent tool of Microsoft—and video clips in my classes to try to engage and capture student attention, and I allow students to use laptops in my classes so that they can be more active. I know some students misuse their laptops, that is, rather than merely take notes they make use of Wi-Fi features to slip onto the Internet to check their e-mail or Facebook pages. Even though I cannot see their screens, I can usually distinguish students who are using their laptops to take notes from those who are multitasking, switching focus back and forth from the classroom to their online accounts. Obviously this distracts them and those around them who can see their screens. Research also confirms that these students commonly do less well in courses.[49] My colleagues and I, at the college level, have changed how we teach in many ways, especially in our assignments. We want students not simply to read and answer questions, but to critically compare and evaluate sources. One must know something before asking a thoughtful question; one cannot seek information and query Google if one does not have some idea that a thing exists before one seeks information about it.

Good News about Information Technologies

Progress in Research

Having easy access to mountains of data alters intellectual life and intellectual activities. Half a century ago libraries were the storehouses of learning. Now that enormous resources have been digitized, modern students don't need to go inside a library but can access vast databases from their dorm rooms. Computers are the essential tool to scan and sift through this vast quantity of data, which no longer can be mastered, much less explored, without digitization.

When desktop computers became available to consumers, I recall that there were discussions about just what computers would be used for in the home. Keep kitchen recipes? Balance a checkbook? Mainframe computers were for computing, and home computers were initially toys. Now computers are a multiskilled tool that are essential for dealing with such large amounts of information rapidly.

Much of the intellectual work of our culture, almost anything that requires writing or gathering of information, now requires the use of the computer.

Some writers find a computer refreshes and fosters creativity in their writing. Salman Rushdie started writing on a computer in 1989. "My writing has got tighter and more concise because I no longer have to perform the mechanical act of retyping endlessly," he explained in interview. "And all the time that was taken up by that mechanical act is freed to think."[50]

The power of computers linked with the Internet for easy transmission to others vastly facilitates cooperative research and collaboration. Take one example from medicine. Computerized pathology slides help doctors make faster and more accurate diagnoses, such as of cancerous breast tissue. Technologies that build on computerized images of biopsies are slowly replacing the glass slides that pathologists traditionally peered at under a microscope. If they have to consult a colleague they wrap up the fragile slide and mail it to another pathologist. Already x-rays are digitized and shared electronically with doctors and stored electronically in patients files. Digital pathology slides require huge number of pixels for the demanding requirements of resolution necessary to enable careful viewing on a computer monitor. Gradually databases can be built up so that computers will play an increased role in analysis, making diagnosis more objective.[51]

So much knowledge has already been gained in areas of science and medicine that further progress will probably occur at the margins of fields. But new technologies also generate so much data that traditional means of analysis are no longer adequate. Consider that a magnetic resonance image of the brain generates images with 50,000 pieces of data for every sequence. To carefully study how a neural impulse moves along a particular part of the brain "is complex and challenging. . . . Modern scientific methods generate too much data to analyze by brute force alone. Without computers to assist in data analysis . . . this would be an overwhelming amount of information. . . . The future of neuroscience is thus inextricably linked to the computer sciences . . . [with] computer simulation and modeling approaches."[52]

But Problems Remain with Information Technologies

Loss of Nondigitized Archives

What is the impact of using a computer on creative writing? To find out, literary researchers would need to compare early drafts with final published work. Anyone who word processes on a computer knows how the delete command so easily sends roughs early to oblivion, unless they are printed out and saved. Born-digital materials, those initially created in electronic form, are complicated and costly to preserve. What scholars once were able to investigate, say the different drafts of James Joyce's *Ulysses*, will probably not be possible for digital poets and writers.

Electronically produced drafts, correspondence, and comments—just a series of zeroes and ones—written on floppy disks, CD's, and hard drives, can degrade much faster than old-fashioned acid-free paper. Even if those storage media survive, the relentless march of technology can mean older equipment and software that can make sense of all those zeroes and ones simply doesn't exist anymore. It's like having a 78 rpm vinyl record but no record player. When our ability to read a document, watch a video, or run a simulation could depend on having a particular version of a program installed on a specific computer platform, the usable life span of a piece of digital content can be less than 10 years. That's a cause for concern when we factor in how much we rely on stored information to maintain our scholarly, legal, and cultural record. "The ephemeral nature of both data formats and storage media threatens our very ability to maintain scientific, legal, and cultural continuity, not on the scale of centuries, but considering the unrelenting pace of technological change, from one decade to the next. . . . For instance, people who created their Ph.D. dissertations in WordStar in the mid-1980s can no longer read them."[53] Though computers have been commonly used for more than two decades, archives from writers who used them are just beginning to make their way into collections.

Other writers discover that publishing their own material online releases their creativity. Last year there were more than 53 million blogs (the word blog is a fusion of two words, We*b* and *log*) on the Internet, with the number doubling approximately every six months. Blogger Andrew Sullivan argues that blogging is a new and revolutionary form of writing. He finds blogging to be the "spontaneous expression of instant thought."[54] For him there is great value in sending out over the Internet his written thoughts quickly, before events have settled and a clear pattern emerges. The vividness and immediacy cannot be rivaled by print, he contends, and he makes a connection with his readers that is unlike what a traditional newspaper might establish with its readers. He finds himself surprised by the expertise of his readers who offer quick comment, correction, and feedback; indeed, he believes his readers provide "a knowledge base that exceeds the research department of any newspaper." Blogs are appreciated for a conversational style in which distinctive character matters as much as authority. In this sensibility, blogging is closer to talk radio or cable news than daily newspapers. Sullivan emphasizes blogging as a collective enterprise as much as an individual one and a new form of authority. Being linked to other blogs is an indication of how central a particular blog is to the online conversation. Another benefit, he claims, is the always-adjusting and evolving collective mind which filters out bad arguments and bad ideas, but he acknowledges the dark side of blogging when there is irrationality and pandering.

There is an even darker side to blogging and much Web information, which is the issue of reliability and veracity. Many blogs and sites like Wikipedia get linked often by the search engine Google. While Wikipedia is the third most visited site for information and current events, and although useful, it simply isn't as reliable as the *Encyclopedia Britannica,* for instance. The issue of reliability and veracity of much Internet information can be summed up by the new term *collective intelligence,* which refers to the collective wisdom of people who use Google and similar search engines. Google makes use of algorithms in its searches in such a way that the more people click on a link that results from entered search terms, the more likely that link will come up in subsequent searches. Because approximately 90 million searches are made on Google each day, this results in a democratization of Web information. Critics like Andrew Keen[55] lament that the ordering of search results reflects what other users have been reading rather than expert judgment.

In short, much Web information is based not on expertise or reliable knowledge, but on least-common-denominator opinion and on what is entertaining. For example, in various aggregation sites, such as Digg.com, which Wikipedia—useful for such up-to-the-minute information like this!—describes as "a website made for people to discover and share content from anywhere on the Internet, by submitting links and stories, and voting and commenting on. Voting stories up and down is the site's cornerstone function, respectively called *digging* and *burying*. Many stories get submitted every day, but only the most *Dugg* stories appear on the front page." Another such site is Reddit.com, which is a social news website on which users can post links to content on the Web. Other users may then vote the posted links down or up, causing them to appear more or less prominently on the Reddit home page. Subscribers can read about a flat-chested English actress rather than a war in the Middle East. When I checked out Reddit I read a mind-numbing discussion of circumcision—confusing because it was based on mere opinion and no factual data. Critics like Andrew Keen deplore that such user-generated content is decimating the ranks of cultural gatekeepers, the trained critics and professional editors, and that the democratization by amateur bloggers and reviewers undermines more thoughtful discourse and belittles expertise.

Plagiarism

Also, the ease with which writers or researchers can scoop up large amounts of data off the Internet can result in plagiarism where the users fail to acknowledge properly either their sources or how completely they relied on the work of others.

Loss of Privacy

Many young people can't seem to live without Tweeting, texting, friending, or posting pictures and even videos online, but social networking carries privacy risks, ranging from embarrassment to identity theft. We live in an era of the viral video when footage of some absorbing slice of life can spread overnight around the globe. Painfully, those on Facebook or YouTube, for example, are learning that it might not be smart to e-mail or post pictures or video clips of themselves at a great party dancing on a table wearing a lampshade. Introverts in real life may enjoy having an online extrovert persona, but do they really want their boss to see this other side? Applicants for jobs are learning that technology has evolved so fast that potential employers, scanning the Internet, may not find those e-mailed photos or video clips as amusing as they seemed the night of the party.[56]

The Internet—particularly social networks—have blurred boundaries so that all relationships are flat, that is, your mother and your best friends and your boss all have access to what is posted. Thus digital walls need to be put up between work and private life, between private information and what one is willing some vast Internet public to know about ourselves. Network users need to realize that posting one's full birth date, or vacation dates, or a child's name leave them vulnerable to cyber criminals.[57]

Other technologies also threaten privacy. Digital cameras in public places and tracking devices like RFID chips can endanger our privacy. An RFID chip, a radio-frequency identification object, is a microchip or tag that is embedded in many everyday items.[58] These tiny chips act as transponders, transmitting and responding to radio signals sent by RFID readers. When the responder receives a certain radio query, it responds by transmitting its unique identification code. The Chicago Transit Authority started using the Chicago Card with RFID tags in 2002, and the Metropolitan Boston Transit Authority has its Charlie Cards enabled with RFID tags since 2006 for entrance to the subway and bus system. When I use my Charlie Card, I simply tap it on the automatic fare-collector, which deducts the fare. I can replenish the amount on my card at a special vending machine. A friend of mine, however, discovered how much data could be collected about his travels when a transit worker was easily able to read off the information and tell him where he had taken the bus over the last months. Highway EZPasses contain RFID tags, as do pet ID tags, and some airline baggage tags.

Bar codes are readily familiar to us from the grocery store, where the cashier runs our purchases over the scanner to the accompanying beep. Although invented in the 1950s, they are now ubiquitous for tracking products and, if we have given our personal information to obtain a rewards card, stores know a lot about when, where, and what we purchase.

Decline of Handwriting

One consequence of the shift from writing to typing to keyboarding and twittering is the decline in the quality of handwriting. After World War II the proliferation of typewriters and phones made handwriting less important. With schools spending less time on penmanship, and keyboarding now so widespread, most people seldom need to write more than a grocery list or sign a check. My father had beautiful handwriting, but I have trouble deciphering my own scribbled grocery lists. My students often resort to printing when taking an exam and some younger students seem to have difficulties deciphering cursive script.[59] Further, in e-mails and text messages, we seem to tolerate grammatical and typing errors that we would not in a hard-copy letter or report.

3

Communication, Entertainment, and Overstimulation

Modern Life Takes Place on Screens

I stop at the bank to withdraw cash from the ATM (with its screen where I can check my balance) and I head for the local chain grocery store. In the store large monitors in the vegetable section and by the checkout show cooking shows and various ads. I use my cell phone to call my wife about what vegetables she wants. If I really needed to, it is possible to take a picture or even a video of what is available and send it to her for her consideration. Once I leave the store I head home to park myself in front of my computer to work—after checking my e-mail or Facebook. After dinner I spend time watching the Red Sox on HDTV or heading out to a movie where powerful images and mighty Dolby surround-sound bombard my eyes and ears with intensity. Or I could just cuddle up with my Kindle to read a thriller.

But my tame experience is nothing compared to the bubble of stimulation that many people immerse themselves in: those who engage with multiple sources of stimuli, such as listening to an iPod while doing homework, or talking on a cell phone while checking e-mail, or texting while attending a meeting, or working on a computer with multiple screens opened.

We live immersed in an intensely stimulating digital environment that conveys waves of information, especially images, enabled by the magic of electronic screens of all sizes and shapes.

This visual culture first appeared somewhere in the transition from modern to postmodern about 50–80 years ago when image began increasingly to prevail over written texts.[1] Certainly, writing and printed text continue to cling to power, but electronic image-transmitting media are pushing aside stodgy print. The early Apple MacIntosh computers began using icons rather than words to summarize a function. Transportation systems adopted graphic maps to guide commuters to the right train. MapQuest, the online provider of directions, offers interactive maps when providing directions.

An image is worth a thousand words, even in Western culture, which customarily has specially valued words, both written and spoken. But these new and extraordinary gadgets give us spellbinding powers to create, manipulate, and reproduce high-quality images. They make us eyewitnesses of a historic transition as one mode of representing reality loses ground and another takes its place without the first disappearing.[2] According to the U.S. National Endowment for the Arts, "Not only is literary reading in America declining rapidly among all groups, but the rate of decline has accelerated, especially among the young."[3] Stephen Jobs, the CEO of Apple Computers, has repeatedly said: "People don't read any more. . . . Forty percent of the people in the U.S. read one book or less last year."[4] Those who do read average a mere six books a year.

The increase in the number of students studying communication and film (with a decline in the study of English and foreign languages) supports the idea that there is an increasing cultural emphasis on images over words. The number of bachelor's degrees in English granted in 2008 represents a decline of 8 percent from 1970; bachelor's degrees in foreign languages have remained about the same. Bachelor's degrees in communications (including journalism, television, etc.) have surged almost 800 percent from 1970, and communication technology bachelor's degrees have increased by almost 1,000 percent.[5] (It is true that some English departments have been blended into communications departments, but again this implies a lesser appreciation of words over images, and flatness or decline in jobs requiring verbal rather than graphic and technological skills.) The number of film schools and film programs established in the last 20 years, for example, has also increased at a notable pace—probably in response to job growth in the cable, video, and Internet film areas. "There's a huge demand for nontraditional visual product," said Kathleen Milnes, a senior vice president with the L.A.-based Entertainment Industry Development Corporation, a not-for-profit group that promotes filming in Los Angeles.[6]

Modern life takes place on screen. "Life-takes-place-on-screen" refers to the presence of computer monitors, high-definition video screens, and smart phones, all of which mediate modern experience for many people. The Nielsen Company's "Television Audience Report" shows that in 2009 more than half the households in the United States had three or more television sets. Indeed, "there continue to be more TVs per home than people—in 2009 the average U.S. home had only 2.5 people vs 2.86 television sets."[7] In addition, television sets are commonly found in bedrooms as well as living rooms, family rooms, kitchens, and children's rooms. This lush world of visual imagery surrounds us, and we have gotten accustomed so quickly to intense, vivid images on our high-definition televisions and on our computer screens that we easily forget that the world was not always filled with such colorful images.

Stimulus Glut

We are embedded, encompassed in a cocoon of information in the form of image and sound. When I walk to the train station for my commute to work, an electronic sign greets me with a message: a repeated voice announcement tells me, "If you see something, say something." I notice many commuters wearing iPods or using cell phones. When I get to work, open up my computer, and first deal with my e-mail, I commonly have 50 messages or more. Occasionally I will journey online and scan YouTube, which usually pulses with moving icons that draw the eye to popular brief videos. As I journey home at the end of the day visual advertisements on the train and on the sides of buses seek my attention. I return home to a pile of mail, much of it junk mail consisting of colorful fliers and catalogues. My high-definition television awaits me with hundreds of channels and on-demand movies. In summary, we live in a very stimulating environment, far more stimulating than at any time in history because of the power and effectiveness of innovative media and communication technologies.

Many of us live amidst stimulus glut. What is that like? Author Neil Gabler uses the example of being in a conversation at a crowded cocktail party and blocking out other conversations to focus directly on the person we are chatting with. But the overabundance of noise and images and information from all sources reaches such a pitch that one needs to shout to be heard above the din. Gabler comments, "In a globalized informational environment where we are awash 24/7 in new bits of data... we are stuck in a kind of virtual cocktail party in which the whole world is present, all talking at the same time, all drowning out one another. In order to get heard, you not only have to shout—you have to shout at the top of your lungs, and even that probably won't do it."[8] All those seeking to contact us and communicate with us and get our attention have to do something so startling as to stop conversation altogether. In effect, one has to blow up a World Trade Center, or paint the streets in human blood, or some other extreme action, to get attention. The global and personal information glut ratchets up the heat, so normal methods of being heard or getting attention don't work very well. This rise in volume helps explain—along with genuine value differences—the shouting on sports radio or the spitefulness of the political talking heads on radio.

This chapter considers some of the changes in recent years that have brought wave after wave of stimulation, entertainment, and information. What are some of the sources of all of this data, this noise, the information glut that surrounds us? This stream of stimuli, this bombardment of our senses, primarily of our eyes but also our ears, renders us a bit numb and resistant to all but the most striking images and insistent sounds.

There is a difference between information itself and the technologies that mediate, communicate, or channel that information. This distinction is important since the medium seems to have become more important than the message. The channel of communication influences the perception of the message; indeed, the dazzling machine or gadget may be more important than the content.[9] For instance, high-definition (HD) television provides crisp, sharp images. If the producers of a news program do not have suitable quality images to accompany a story, another, lesser story gets air time because images trump words. I find myself distracted from a story if a newscaster has skin blemishes that the brutally demanding HD cameras display. The quality of the image does detract from the message. But more on this below.

Cell Phones and Texting

Smart phones, iPods, and computer tablets like iPads, are useful and ubiquitous. Many of these portable handheld devices enable their owners to capture videos and send them to others, to watch movies and live sports, and so forth. A MacArthur Foundation Report on Digital Media uses the phrase "always-on communication" to describe how young people use electronic media to maintain friendships and romantic relationships as well as to hang out with each other as much and as often as possible.[10] This sense of being always on and engaged with one's peers involves a variety of practices, varying from browsing through extended peer networks such as MySpace and Facebook profiles to more intense, ongoing exchanges among close friends and romantic partners. Youth use MySpace, Facebook, and IMs to post status updates and become, in turn, exposed to new standards and norms for participation in specialized or collaborative communities. These unique channels for communicating have altered many of the conditions of socializing for youth, even as they build on existing youth practices of hanging out, flirting, and pursuing hobbies and interests.

Sometimes this high level of connectivity can get in the way of other responsibilities. To take one example, complaints have risen sharply about young lifeguards who were texting on the job. Some of these college students did not want to feel disconnected from their gadgets and their friends even if their job was to be vigilant for signs of trouble at the shore. A 45-year-old man drowned at a lakeside beach in Illinois while the young guard was texting, according to witnesses at a civil suit.[11]

Cell phones are now indispensible for teen and digital-native communication, and texting via cell phone has become the preferred means of basic communication among teens and their friends, with cell phone calling is in second place, as reported by the Pew Research Center's Internet and American Life Project.[12] Seventy-five percent of Americans aged 12–17 now have cell phones, up from 45 percent in 2004. I give emphasis to teens and digital natives because they tend to be early adapters of new media—if they can afford it—and trends established among these early adapters tend to gather force and percolate up to their elders.

Teens love their cell phones, especially the messaging features. Only 30 percent of teens use traditional land lines, 25 percent use social network sites, 38 percent call on cell phones, 54 percent of teens text message daily, while 33 percent talk face-to-face. The volume of texting is startling. One in three teens sends more than 100 texts messages a day, or 3,000 a month. I confirmed this in interviews, and have before me the inch-thick bill for one bright, attractive junior in high school who texted 4,280 times in February of last year. Calling by cell phones is still a central function for teens, especially communicating with parents. Girls (86% per day) are more likely than boys (64%) to use both text messaging and voice calling. Teens typically make or receive at least five calls a day, with texting almost double the amount of calls.[13] Thus people seem to have plenty of connectivity, but are actually talking less to each other.

One college student of mine said about her steady texting that she feels much better because someone knows all the time where she is. Another said that each time she observed something interesting, she had to text someone to make the experience more solid. One high school girl shared with me that at night she commonly texted from 9:00–11:00 P.M., and then, in a panic, would begin her homework and end her day by taking her phone to bed; her very last action at night was to text her best friend.

To be sure, cell phones are invaluable to coordinate meetings, to arrange for pick-ups at train stations, or to get help from a spouse about choices for dinner while at the grocery store. When riding the T or commuter train I commonly observe skilled texters using their thumbs to send messages, and will often overhear conversations, sometimes intimate, sometimes banal: "Hi, it's me. What are you doing? Me? Nothing. What about tonight?" Or: "I'm on the T. Three stops away. See you in five minutes."

Many feel the cell phone enhances the safety of users and keeps them connected. The vast majority of parents and teens agree that they feel reassured because parents and children can reach each other no matter where the teens are. Eighty-four percent of teen cell owners agree that they like how their phone makes it easy to change plans quickly. This extreme flexibility can be also a problem, as we shall see in chapter 5.

Cell phones, with expanded functions and new applications (apps), have become much more than phones: they now become video recording devices, pocket-sized Internet-connected computers, portable game players, music players, and digital cameras that can transmit pictures. In my health club I was surprised to see a man strap his high-end phone on his arm, even though cell phones were not allowed, in the locker room. "I use it to listen to music while I work out," he explained patiently, dismissing any concerns health clubs have about privacy or inappropriate picture-taking. Indeed, high school students report that they have sent nude or nearly nude images to someone (about 4% report this "sexting" phenomenon), and one in three teens (34%) say they have texted while driving, which works out to about a quarter of all American teens of the ages 16–17.[14]

One consequence of the ubiquity of cell phones is the decline of public pay phones. Even in busy locations, such as outside the courthouse in Queens, New York, there are 50 percent fewer public phones than there were just a few years ago, as reported by the New York Department of Information Technology and Communication.[15] Another consequence of the utility of cell phones is the decline of home land lines, with about a quarter of American households using only cell phones.[16]

Some cell phone users exhibit a compulsion that resembles addiction, that is, they feel they must respond to their phones, even though the situation is inappropriate. In class, for example, I have asked students not to use their phones, especially for texting. In spite of penalties, some students have texted in class or left the room to respond to some message or text; when asked about their behavior, they have admitted to me that they are "kind of addicted" to their phones. The impulse to respond to immediate stimulation can provide the brain with a squirt of dopamine, a kind of reward that makes it hard *not* to use the phone. The stimulation provokes excitement, and in its absence, people can feel bored.[17] In an Apple Store recently, one of the young sales associates, even while talking with me, simultaneously was playing a game on her iPhone. I could only regard her as providing me with split attention and she needed the buzz of further stimulation that the iPhone app provided her. Sherry Turkle, the MIT researcher, speaks of the greater force compelling users to keep checking their screen: "There's something that's so engrossing about the kind of interactions people do with screens that they wall out the world," Turkle says.[18] Turkle goes on to say she has interviewed children trying to get their parents to stop texting and they get resistance: "Oh, just one, just one more quick one, honey."[19] The image of a toddler trying to get his mother's attention as she is absorbed in the BlackBerry is sobering and compelling.

Television and Its Advertisements

Television screens were the first incursion, seemingly benign, of the modern visual revolution into American homes. However, we have not yet fully integrated this revolution into our daily lives. Even though television has been a powerful cultural presence for more than 60 years, we still have not managed those glowing screens, and now new types of screens have arrived, such as those found on computer monitors and Personal Digital Assistants like BlackBerrys and iPods and smart phones.

A major component of viewing commercial television is watching advertisements, which very often are clever and strikingly designed to break through the clutter of all other visual stimuli. After all, the purpose of a television program is to set the viewer up for the advertisers' pitch. They are paying the networks to assemble an audience for them to appeal to.

However, watching television teaches one to be distracted and develop a shortened attention span. Movement catches our eye, and most television commercials are rapid, punchy, eye-catching, with scene cuts often lasting just a second. Thus a 25-second commercial might present its message through 20 or more different images flashing on the screen. By contrast, a scene from a 1930s black-and-white movie or from a television commercial of the 1950s may last for a minute or a minute and a half, with each scene lasting for 10, 15, or even 20 seconds. The editing of these old films and ads make them feel slow and boring to us who have grown used to rapid shifts from one image to the next.

Television clutter refers to all the commercial time, promotions, and upcoming program teasers that precede, follow, and even interrupt the program itself as a pop-up either at the bottom or corner of the screen. "In the primetime slot, non-programming time on network television was 16:43 minutes per hour. The daytime level of advertising was 20:53 minutes per hour. Network news showed 18:53 minutes of commercials per hour and late night news aired 19:06 minutes of ads per hour. The most 'cluttered' program in all of TV, according to the report, was ABC's *Good Morning America*. . . . . On cable, the Fox Family Channel was the most cluttered with 18:03 minutes of commercials per hour; E! came in second with 17:19 minutes of ads per hour; and MTV was third with 17:19 minutes per hour of clutter."[20]

In order to attract our attention through use of computer graphics, many ads display striking and humorous nonrealities—such as a woman being pulled by her coffee mug as if by powerful force or a ceiling falling in on frat boys drinking their favorite beer while they neither get injured nor even notice the nonreality. A recent Honda ad depicts giant robotic mosquitoes that suck the life out of trucks and cars; it is a clever, attention-grabbing ad but likely would terrify young children, who still do not have a firm grasp on conventional reality. All this might be amusing, perhaps, but is another example of the digital ability to blur the everyday world with a fantasy world. More on this below.

Once saved or recorded sounds and images could be transmitted though the airways—via radio and then television—the information revolution accelerated. Television sets became common in the United States after World War II. It was only 50 years or so ago (a very short time in human evolution) that most TV programs used to be live. Later, networks introduced videotape for delayed broadcast. It was in 1956 on the Jonathan Winters comedic variety show that the experimental process of videotaped performance was introduced. In this instance, a prerecorded performance of Dorothy Collins singing a two-and-a-half minute song was featured within the live show.[21]

What If There Were No Television?

What would happen if there were no television? One study documented the effects of television's arrival on a small Canadian town "Notel" that had been without television reception for a decade into the television era. Researchers studied the television-free children and families, comparing them with the population of two demographically similar towns, one of which had only one TV channel and another that had many. The findings were dramatic. Before television, "Notel" children tested significantly higher on creativity, reading, and imaginative skills than the kids of other towns.

After television came, the researchers retested the children and found their scores were down, now equal to other kids served by television.

The researchers stress that it is *not the act* of television itself or what people watched that lowered scores, but that television watching *displaced* other more valuable experiences, such as reading, playing, interacting with one's peers. Creativity, reading skills, and high participation in other leisure activities are diminished when TV arrives in children's homes.[22]

Enhancing Ads

Innovations in computer graphics and media technologies have transformed advertisements and made them more effective—or intrusive. It has been said that the average American is exposed to at least 3,000 ads everyday, perhaps as many as 5,000. One market research firm estimates that a person living in a city 30 years ago saw up to 2,000 ad messages a day, compared with up to 5,000 today. "What all marketers are dealing with is an absolute sensory overload," says Gretchen Hofmann, executive vice president of marketing and sales at Universal Orlando Resort. The landscape is "overly saturated" as companies press harder to make their products stand out, she believes.[23]

Because of a class I teach on visual literacy I've looked closely at how commercials work. Many of the most clever flicker with an image each second, a rapid visual stimulation that draws in our eye. (The term is "orienting response," the tendency of our eye and attention to be directed toward movement; it is probably a survival mechanism that worked well for our ancestors several million years ago in their vigilance against predators.) The best ads tell an engaging brief story within 20 or 30 seconds. But routinely this ad is followed by another ad. And then another. And another. To break through the carapace of a jaded, passive viewer, ads need to be either brilliant or numbingly repetitious. As I watched the U.S. Tennis Open on television, several ads kept being repeated, and this bludgeoning was their power. Advertisements are not new, but the jazzy, punchy ads of contemporary television are new. Television is a clever technology in the service of commerce, with the task of gathering an audience for the advertisers. Probably some of our culture's cleverest graphics and video work is on display in television ads.

Blatant advertising is just one contributing factor of our stimulus-saturated environment. The feeling of ubiquity may also be fueled by spam e-mail messages and the increasing use of name-brand items in TV shows and movies, a trend known as product placement. Old-style billboards are being converted to digital screens, which flash public service messages and ads. Those monitors in stores, gyms, doctors' offices, and on the sides of buildings are part of this trend toward ever more intrusive marketing and the barrage of irrelevant messages.[24]

In this hyperstimulating environment, the entertainment industry and advertisers in particular must ratchet up even further the intensity and novelty of images to cut through the image glut.[25] (Literally, *advertisement* means "pay attention to, turn one's attention to.") Thus it is not only the number of images and sounds, but their quality as well. An old-fashioned analog television could not match the sharpness and intensity of images on a high-definition television screen or computer screen or iPad. The backlit liquid crystal display screen gives images a compelling luminosity impossible on paper or older screens. And now on our video and computer screens are the clever and innovative graphics of modern video commercials. Digital cameras of 8 or 10 or 12 mega pixels, utilizing PhotoShop software, allow the average person to create compelling images. Tiny iPods allow listeners to access and download and carry around thousands of their favorite songs with remarkable fidelity. Oh, yes, and smart phones allow one to capture video or images of whatever is happening around oneself and transmit streams of images to friends.

So Many Images, So Vivid, So Different

Pictures of all kinds, whether on the screen or reproduced on paper, not only nowadays demonstrate images of very high quality, but the intensity of the images and their colors have very much improved. High-definition television, of course, and digital cameras make available images with a sharpness and saturation of color not formerly available. High-definition televisions have the ability to render spectacular clarity and detail, because they have more pixels. The older standard definition carried far less visual information, and appears blurred and dull in contrast to the clarity and brightness of high-definition screens. Because color technology in photography and film is comparatively new, some might think of the past in terms of the black-and-white images of old movies and old snapshots. Of course, as we see restored World War II film footage that was in color we are emphatically reminded that reality has always been in living color, even though some clothing might have been somewhat drab because of natural fibers and natural dyes. We have a greater range of colors available to us—not only in electronic media, but in clothing and even in the paint that is on our walls. Paint manufacturers advertise a rainbow of colors and clothing manufactures offer an extraordinary range of colors. For example, on a recent tour of the Behr Paint website, I came across at least 147 shades of the 7 or 8 basic colors; Behr pamphlets describing their interior and exterior paints offer seemingly unlimited possibilities for flats, satin enamels, semigloss enamels, hi-gloss enamels, and porch and floor paints.[26]

Not only has the quality and quantity of images increased, the content has changed and intensified. Over the past half-century, there are more graphic sexual and violent images, images that were culturally repressed or simply chosen less frequently. For example, during World War II the premier media outlet for images, *Life* magazine, with rare exceptions did not show pictures of dead American soldiers.[27] Now violent images are common in movies and on television, and the Internet has provided an outlet for the easy availability of pornography and an extraordinary increase of it. Pornography has always been at hand but much of it was narrative, text and words, and legal barriers prevented wide distribution. In 1973 there were fewer than a thousand "adult movie" theaters, the main outlet for visual pornography. Then came the VCR and DVD's—one in five videos became porn, which facilitated viewing erotic movies at home on tape, and now also via the Internet.[28] Even television commercials push into new territory. With Viagra or Cialis ads now well established, it is hard to think of anything embarrassing in this age of erectile dysfunction ads, sanitary product ads, or ads for diarrhea or overactive bladders. Network sitcoms still seek a broad appeal now, even though writers can no longer rely on shared values or widely accepted norms of inhibition. Comedians try to stretch the limits of humor by becoming ever more lewd and offensive. Once cultural sensitivities harden and taboos are vanquished, it is harder to find fresh forms of adult-only material. The only way to ratchet up attention-grabbing stimulation is to go for more shock, ever more explicit images, and use production techniques of quick cutting and fast pacing to hold our jaded attention.[29]

Altering or Enhancing Reality

But it is not just visual stimuli, it is also the manipulated and transformed nature of the images that attracts my attention—and my concern. Again a distinction: the electronic manipulation of televised images is different from the computer-assisted creation of totally new images or graphics.

Consider sitting down and watching a baseball game with me, the 2010 World Series. Digital technology alters how we see this most traditional sport as television presents it, by jazzing up the 150-year-old sport with extra cameras and replays so that leisurely old baseball is almost unrecognizable. The batter approaches the plate. Oops, we get a quick replay of him hitting a home run from a previous game. Strike three. He's gone. Before the next batter steps up, we get a quick replay of the final strike. The next batter hits a sharp grounder between the third baseman and shortstop. We get three quick views of the hit. One's sense of time speeds up and one's sense of reality is stretched and garbled, since it is not always immediately clear if the images are replays or live action. If the inning is over, during the commercial break, we get several visually interesting ads, as well as a teaser commercial for an upcoming network comedy show. Then back to the game, where we might see another replay of the last out, as if our attention has not been splintered enough. Our minds quickly adjust to these patterns, and we become visually literate in the style of the modern media—and it is difficult to go back to the slower and, frankly, duller mode of earlier films and television. We have been changed.

The difference is highlighted by the accidental discovery of a piece of the past: a pristine black-and-white recording of the televised deciding game of the 1960 World Series. This kinescope recording (an early relative of the DVR) is the only complete copy of a game considered one of the best baseball games ever played. Author Richard Sandomir comments, "The production is simple by today's standards. NBC appeared to use about five cameras. The graphics were simple (the players' names and little else) and rarely used. There were no instant replays, no isolated cameras, no analysis, no dugout reporters and no sponsored trivia quizzes."[30]

Secondly, in addition to manipulated images of reality, there are the remarkable images created by computer graphics, the CGI (computer-generated imagery). I have mentioned above the advertisements crafted with the best of computer-aided graphics, which enables digital images to be transformed, morphing from an image of one object into something else. As in the film *Matrix,* or another science fiction movie like *Avatar,* or in cutting-edge ads, CGI is often the transforming power behind the special effects in films, television programs, and televised commercials.

Computer-generated imagery is used for visual effects because computer-generated effects are more controllable and less expensive than other more physically based processes, such as constructing miniatures or hiring extras for crowd scenes. Also, CGI allows the creation of images that would not be feasible using any other technology. Wikipedia describes the use of CGI in computer games and in simulators. Simulators, especially flight simulators, make extensive use of CGI techniques for representing the outside world. Such graphics subtly distort our sense of visual reality and make us ask the question, "What is real?" In a digital age, an age that is increasingly visual rather than verbal, a media age rather than a text age, what is real is what can be seen. How often do we look at an image now on television where the four letters "Live" are in the upper left-hand corner? But even when the word is used the scene may or may not be broadcast live. Because of the superb technical qualities of modern recording devices it could be a replay of something that was live a few minutes ago or a few weeks ago. Several weeks ago I showed my students a picture of a woman in a wedding dress in flames. It's a horrifying image that is baffling because the woman in a wedding dress smiles as her dress is burning fiercely. One can only look in horror. Students were shocked at the image but were certain that the flames were digitally added to the smiling woman's white dress. Because students have grown accustomed to images in ads and films that have been digitally altered they presumed the burning dress was "not real"—but in this case it was, as the website (altf.com) from which I downloaded the picture explained the intentions of the photographer and even added the caption, "Mothers should close their eyes." Students have acclimated to a visual age, and have grown accustomed to a digitally altered reality, whether it is in films or the many advertisements that mangle our sense of reality—cars that ride upside down, actors that change into other creatures.

Computer-generated imagery can aid in teaching and learning and can reduce expenses. For example, medical schools have found a way to use such imagery in the anatomy curriculum, thereby making it unnecessary for medical students to spend so much time dissecting cadavers. Increasingly, computers and their software can show things that lifeless cadavers cannot, such as blood pumping through the heart chamber. Incidentally, the software is based on careful CT scans and MRIs that allow users to study images of the body from every angle and on every plane. This permits students to peel away muscle on a virtual leg to see the bone beneath, then reattach muscle and push a button to see how the leg moves. (As to the fabricated heart, one might wonder, perhaps, if something gets lost by a medical student not holding a real human heart, that symbol of altruism from a stranger, the gift of a heart for science?)[31]

Shift toward the Interactive

As in so many other areas, young people are at the forefront in their use of many of the communication and information technologies. The trend is unmistakably toward being interactive; that is, interacting with the medium that one is viewing or using. Young people, the digital natives, live in a media-rich environment. They spend considerable time using digital and electronic technologies, not only in their studies but also in many other areas of their lives: to communicate, to state and seek opinions, to socialize, to remain informed and to live day-to-day. For their entertainment, natives are more apt to use online programs; they spend almost 50 percent more time than immigrants participating in Facebook and My Space and microblogging on Twitter. Thus, depending on the type of technology used, they are more exposed to ads. They are moving away from television.

Television is a one-way broadcast medium, a passive channel of information. The Internet, by contrast, allows collaboration and invites simultaneous participation from multiple users. The Barack Obama campaign benefited from the staff's digital prowess, and they took advantage of both

Facebook and MySpace.[32] In the late 1990s the average American 18-year-old watched almost four hours of television a day, about 22.4 hours per week. The 80 million digital natives born between 1980 and 2000, who grew up with the Internet, watch about 17.4 hours a week, less than the generation before them, but the digital natives or "NetGeners" spend almost twice that amount of time on the Internet. Young Americans have switched to the computer screen and the Internet for much of their entertainment.[33] Informal surveys of my 80 freshmen students confirm this. They informed me that the average amount of time they spent at their computers surfing the Web each day was commonly three and a half hours. Four of them were extreme users, in my judgment, logging in with eight or more hours per day. One young man put it this way: "Sure, I watch some television, but really spend the bulk of my time on the Internet, surfing some of my favorite sites, such as YouTube and CollegeHumor.com."

Whatever their entertainment qualities or distracting value, such websites offer students and bored workers a "four-minute vacation from work spreadsheets or school term papers . . . sites for videos of wedding bloopers, photos of innuendo-laden billboards and college humor . . . videos of cats playing piano."[34] Last year there were 25 million new videos posted on sites such as YouTube, many of a humorous or silly nature, with 25 billion visits. But there are also videos of deadly firefights and bombings uploaded by young soldiers serving in Iraq and Afganistan.[35] YouTube, another favorite site, plays a very different role as part of the college athlete recruitment process. Haverford College's lacrosse coach, Colin Bathory, said of the use of the video exchange site: "YouTube is taking over. And that's a good thing. Too many kids had to pay for expensive DVDs. For players to be able to shoot a game video one morning and for me to be able to see it that evening, that's a breakthrough."[36]

Attention and Attention Overload

What information consumes is the attention of its recipients. But a wealth of information creates a poverty of attention. Our attention is a precious commodity, easily captured by irrelevant stimuli. In an environment that is visually exciting and crafted to catch our attention, it is not easy to remain focused.[37] Indeed the digital environment seems to lead to feelings of information overload. Many users of the Internet now gravitate toward sites that are designed to filter and organize information into manageable levels. Such sites as Digg, Reddit, Delicious, or Mixx rely on the wisdom of crowds to determine which stories are worth spending time on. The bottom line, however, is that the content streaming through these websites throughout the day tends to be a hodgepodge of random articles and breaking news, which contributes to the sense of even more data noise and more overload. Traditionally gatekeepers like the trained staff of the *New York Times* or *Washington Post* or *CBS News* or *Time* could be entrusted to perform this filtering and evaluative function—but the dam has broken and the greater flood of information seems bloated and frequently unmanageable.[38]

Television tries to pull its viewers into more interactivity by appealing to its audiences to vote for a favorite performer on *American Idol* or to send in news tips or news photos. *America's Funniest Home Videos* long ago (by television standards) began to profit from the contributions of viewers who send in video clips for financial rewards. But television can't match the speed and virtual instantaneity of YouTube and other Internet sites that rely on input from users. The millions of blogs exemplify the desire of many to contribute their thoughts and opinions. Technorati, the online site that indexes and tracks blogs, estimates that some 175,000 blogs are created daily, and that in cyberspace there exist 113 million blogs, with 7.5 million of them active. Meanwhile, 184 million bloggers are creating 570,000 posts every 24 hours, reaching 70 percent of Web surfers daily.[39] (The name Technorati appears to be a blend of the words technology and literati, thus suggesting a blend of technological intellectualism.)

The Ultimate Interaction: Video Games and Online Games and Poker

Let me turn to an area that is the heavyweight weight champion of digital interactivity, and that is gaming: video and online interactive games. Pong and SuperMario are the dark ages of video games measured by sophistication of graphics and tasks. Modern computer and online games display excellent graphics, engaging challenges, and the promise of virtual immersion in an alternative world.

One phenomenon stands out, *World of Warcraft,* often shortened to WoW, a massive multiplayer online role-playing game (MMORPG)[40] put out by Blizzard Entertainment in November of 2004. *World of Warcraft* is one of a series of games set in the fantasy Warcraft Universe, which was introduced in 1994. *World of Warcraft* takes place in the world of Azeroth, four years after the events of the previous game, *Warcraft III: The Frozen Throne.* There have been three expansion sets so far over the last three years: the *Burning Crusade, Wrath of the Lich King,* and *Cataclysm.* As of December 2008, WoW attracted more than 11.5 million monthly subscriptions, making WoW the world's most-subscribed MMORPG, retaining close to two-thirds of the multiplayer-online-game subscribers.[41]

I have used websites such as *World of Warcraft,* gamespot.com, blizzard.com, and Wikipedia[42] to gain some entrance into this new and strange cyber world with its own jargon and rules. To enter as a player, one must pay for a subscription that enables one to control a character within a game world, explore the landscape, fight various monsters, complete quests, and interact with others. A player must select a realm. Realms fall into one of four categories. Realms are either *Player versus Player* (PvP), where combat among players is common, or *Player versus Environment* (PvE), where the gameplay is more focused on defeating monsters and completing quests. *Roleplay* (RP, RP-PVP) offers variations of the other realms.

To create a new character, in keeping with the storyline in the previous games, players must choose between the opposing factions of Alliance or Horde. Characters from the opposing factions can communicate, but only members of the same faction can speak, e-mail, and share guilds. The player selects the new character's race (species), such as Orcs or Trolls for the Horde or Humans or Dwarves for the Alliance. Players must also select the class for the character, with choices such as mages, warriors, and priests available. As characters become more developed, they gain talents and skills, requiring the player to further define the abilities of that character. They may learn professions such as tailoring, blacksmithing, and mining, as well as secondary skills, such as cooking, fishing, and first aid. Characters may also form or join guilds, which allows characters in the same guild to share a guild name, establish special communications, and so forth. Much of *World of Warcraft* play involves *questing,* also called *tasks* or *missions.* It is by such quests that much of the game's story is told.

Quests commonly involve killing a number of creatures, gathering a certain number of resources, finding a difficult-to-locate object, speaking to various nonplaying-characters (NPCs), visiting specific locations, interacting with objects in the world, or delivering an item from one place to another.

Online *Poker* is quite different, but for its players perhaps equally addictive as *World of Warcraft*. A Google search for "online poker" yielded 4,290,000 results in .30 seconds. Online poker is the card game poker played over the Internet. There are specialized websites that will provide guidance and rankings and up-to-the minute information on the eGaming industry and worldwide gambling, such as Christiansen Capital Advisors (CCA) and Global Betting and Gaming Consultants GBGC. Indeed, CCA produce a set of revenue estimates and projections by game, geographic region, and year. They issue *Insight: The Journal of the North American Gaming Industry*.[43] The estimate of online poker revenues was $82.7 million for 2001 and will be over $24 billion in 2010. The CCA has been partly responsible for a dramatic increase in the number of poker players worldwide.

Some commentators make great claims for the value and merits of video and online games. It is true that some games can lead to players becoming affiliated with new groups of people. Perhaps the three points of our nation's IQ gain each decade since World War II reflect the spatial skills cultivated by video games. Perhaps the new connections and new pattern recognition skills prepare us and our children for future learning and future problem solving. It does seem true that the users of such games are learning to customize everything, want to have fun, believe that speed in technology and in everything else is normal, and regard constant innovation as a fact of life.[44]

Do Virtual Reality and Computer Graphics Blur Reality?

Of course, fantasy can be enlivening and entertaining, and since ancient times people have enjoyed immersing themselves thoroughly in a fantasy world. This also holds true of more modern people such as the readers of the Sherlock Holmes stories or Harry Potter fans. But reading allows one to pace the experience and exercise some control over it. The power of virtual world technologies to immerse the gamer or participant in alternative worlds, by contrast, is enormous and unprecedented.

There is a difference between direct experience and virtual experience. Also, there are different degrees of virtual experience. We can be drawn into a play or movie or novel and share the emotions of the protagonists. Modern image-producing media has power to seize our attention and feelings to a remarkable degree. The immense grief and sense of loss that millions felt at the death of Princess Diana and of John F. Kennedy represents a new inability to distinguish between the virtual world of media experience and the real world of direct experience.[45] And one can wonder what the fans and spectators at sporting events are seeking to achieve by frantically waving at the television cameras panning over the crowds. Considering how popular social media like Facebook and MySpace are, does being on screen, be it a television or computer screen, enhance somehow one's sense of existence? Are people living *through* media in some new and different way?

The term *virtual reality* is commonly used to refer to computer-simulated, immersive 3-D environments that can imitate places in the real world as well as in imaginary worlds. Most current virtual reality environments are primarily based on visual creations that are displayed either on computer screens or within a special helmet with stereoscopic display terminals, with special headphones to provide stereophonic sound. Thus, users or players can interact with a virtual environment either through the use of standard input devices such as a keyboard and mouse, or multimodal devices such as a wired glove. The simulated environment or virtual world can seem similar to the real world in which pilots are trained for combat, or it can differ significantly from reality, as in some virtual reality games. Most who seek out virtual reality games are seeking a different experience of reality—a drug-free heightened sensory experience.

Virtual reality pioneer Jaron Lanier coined the term *virtual reality* and in the late 1980s he led the team that developed the first implementations of multiperson virtual worlds using head mounted displays, for both local and wide area networks, as well as the first *avatars,* or representations of users within such systems. Lanier founded a research company where he and his colleagues developed the first implementations of virtual reality applications in surgical simulation, virtual sets for television production, and assorted other areas.[46] I dwell on this because some companies are developing this technology, which enables users who are at different distant geographical sites to collaborate in real time in a shared, simulated environment as if they were in the same physical location. This radical new paradigm for human-computer interaction brings together networking and media technologies to move meetings into a virtual world that feels real. If Photoshop could alter pictures (make a fat person seem thin, for instance) and blur reality, the application of technologies out of the game realm into the work realm seems a further blurring of reality and fantasy.

Byron Reeves and J. Leighton Read in their book, *Total Engagement: Using Games and Virtual Worlds to Change the Way People Work and Businesses Compete,*[47] argue for the introduction of these digital resources into the work space to transform dull aspects of work and make it special. If the virtual space has special properties, why not make use of those qualities? Because computers allow us to do things otherwise impossible in the real world, why not use these possibilities? For instance, using gaming technologies allows a player "to quickly change visual perspectives. With the press of a key she can view the world from the eyes of her character or put the 'camera' anywhere she pleases, such as looking down from above or from over the shoulder of her character so she can see her position in space. And she's not limited to just walking: flying, teleporting, or catching a ride on a dragon are also options. None of these are real world possibilities, but it's also true that all are easy to understand because they incorporate familiar features of reality" (pp. 67–68).

The software programs, scanning devices, specialized gloves, and robotic arms of companies like Polhemus have extraordinary applications in the world of art, art restoration, simulation trainings, and manufacturing. For example, to rescue decaying stone carvings in Asia, a team using a Polhemus scanner (a handheld, portable 3D scanner creating real-time images) and a laptop computer scanned several ancient stone carvings, which enabled museum conservators to produce an exact reproduction of the carvings. In a different use of these technologies, an artist can scan a delicate wood sculpture, and save the file in various formats that can then be imported into a CAD (computer-aided design program) or modeling software in order to create copies.[48]

Areas of Noise Stimulation

Amidst the discussion of visual stimulation, I don't want to neglect the role of increased aural stimulation, that is, sound. Not all sound is noise.

Many people who make use of iPods or MP3 devices or cell-phones don't regard music or phone calls or recorded messages or uploaded programs as noise. Certainly it is data, with the same ones and zeroes as visual imagery. One among modern experiences, thanks to technologies of amplification, is attending a pop music concert that seeks to immerse the audience in the intense sensory experience. Amplifiers saturate one's ears and the pyrotechnics flood one's eyes. The slam-bang blockbuster films on a large screen with Dolby surround-sound aim to embrace the audience's attention and provide a strong sensory experience, if not an emotional one.

Other than this aural data (sound) there is noise. There is good evidence that noise pollution is increasing. Of course, not all aural data (sound) is noise and not all noise is bad. Transportation vehicles are the worst offenders, with aircraft, railroad stock, trucks, buses, automobiles, and motorcycles all producing excessive noise. Construction equipment such as jackhammers and bulldozers also produce substantial noise pollution. Researchers seek to raise awareness that noise has negative health consequences. Indeed, there are sites where one can download white noise for one's MP3.[49] In Europe there is such concern over the rise of noise—especially in urban areas—that governments are now mandating "noise maps" to assess and manage environmental noise.[50]

Vignette: I'm standing on a Boston MBTA platform waiting for a train. There is a new electronic sign giving the time and occasionally electronic messages. Then a public announcement blares forth—not to announce a train or a delay—about helping the Boston Fire Department keep our subways safe from fires in stations and even an announcement in February reminding riders to get their flu shots. I ponder just how helpful or important this kind of announcement is. Or does it just add to the level of irrelevant or useless data?

Low-energy light-emitting diode (LED) technology allows for large electronic billboards with various commercial and public service messages. Similarly, nonaural electronic signs appear overhead on Boston's ring-road reminding us, "Fasten Seat Belts." Of course, sometimes these signs do announce a breakdown ahead. But one wonders if the distraction is worse than the occasional useful notices.

By Contrast

How did we get here? How did American culture burgeon into such a overwhelming, exciting, bewildering mash-up of sensory experiences?

The world used to be very different. Let us imagine, for the sake of comparison, a couple in Europe four or five centuries ago living as peasants in dwellings near their fields. During their lives most never would have journeyed more than 10 or 12 miles from where they were born; there was little reason to leave home unless they were going off to war, perhaps the Crusades, for instance, or a pilgrimage. The largest manmade object they would ever see was a castle or perhaps a cathedral if they had been able to walk to a city. To stand inside a light-filled cathedral with stained glass and singing would have been stimulating and otherworldly. They knew only a limited number of people within walking distance of their village and farm. Time moved predictably, marked by the rise and setting of the sun and the rhythm of the seasons. Perhaps they heard the church bells ring the Angelus three times a day. Nobody needed to be on time, since there were few if any clocks. The fastest they could travel would be to run or ride on a galloping horse. The fastest thing they might have seen would have been the flight of an arrow or a flash of lightening.[51] Few had ever seen a book, much less read one. In short, they did not have media, whether books or electronic screens, to mediate between them and reality.

But we do not have to journey far back in time to note the rise in media stimulation. Those of us in middle age can remember pre-TV or black-and-white television.

We can mark the beginning of the information revolution, not with the invention of the printing press, but with the development of technologies that record and preserve sounds and images. First of all came the invention of photography. It took a while but eventually photographers displaced painters as the primary makers of images and indirectly caused the rise of artistic movements like impressionism. Why should a painter strive for realism when the camera could do a better job? And this was long before the rise of color photography. Techniques improved for the easy reproduction of visual images, such as prints and lithographs, which provided the masses with continuous access to the symbols and icons of their culture. Prior to this graphics revolution most people saw relatively few images. In churches they could see colored windows depicting Jesus and the saints; images of national leaders would be seen on coins and mainly in homes of the wealthy or in government buildings. Books contained images, but books were expensive and in the hands of relatively few people. In short, visual images were not a conspicuous part of the environment, and this scarcity gave them more impact.[52]

A Consequence of Too Much Information: Distraction

A negative side of an abundance of information is that it is not easy to differentiate between trivia and material that is important for us. Our attention can be captured by irrelevant stimuli and so we need to work harder to keep our attention focused.

On October 9, 2009, two Northwest Airlines pilots failed to pay continuous attention to their instruments during a flight and overshot their destination by 150 miles. Pilots Cheney and Cole told investigators they lost track of time while using their personal laptop computers to look at new airline scheduling software related to Northwest's merger with Delta Air Lines. They were piloting the Airbus A320 from San Diego, California, to Minneapolis, Minnesota. More amusingly, a burglar in West Virginia who stole diamonds then paused to check his Facebook page before leaving, and his forgetting to sign off enabled the police to nab him. Needless to say, the encroachments of stimulating gadgetry and wireless cell phones in cars causes more attention failures.[53] In hospital and medical contexts, noise levels are rising; distractions and interruptions can cause nurses and doctors to sometimes lose focus. If this is in a situation where drugs are being dispensed, the safety issues are obvious.[54]

Amidst information saturation, television works very hard to grab the attention of viewers. "Breaking news" banners, logos, promos of upcoming stories and stock market quotes, all appear simultaneously on screen. Because screens began to be crowed with too much information, one of the cable news providers, CNN, finally decided to reduce some of the "crawl" or unending ribbon of news bits flowing relentlessly across the bottom of the cable news programs.[55] This was a tacit rejection of the information overload that has typified television news for much of the past 10 years. That scrolling ribbon of news at the bottom of the screen is the cousin of the stock market ticker that began scrolling across cable business news in the 1980s and is also related to the flashing news headlines that circled the headquarters of the *New York Times* as far back as 1928.

Sports, which television communicates well, can become a ludicrous display of useless statistics and diverting distractions. When the football announcer tells viewers that the New York Jets have won only two games in which they were four points ahead with less than six minutes to play, does this add to my useable knowledge? At its best, television can powerfully convey information, but its mode of communication is to present content—be it news, religion, weather, or history—*as stimulating entertainment.* By its very nature as a medium that renders the viewer passive, television has to jazz up its presentation to keep viewers alert and paying attention. The result trivializes serious discourse and renders most communication a form of entertainment that allows little time for reflection or criticism. Neil Postman, a fierce critic of television who also acknowledges its power to convey information effectively, points out how easily television moves from a solemn topic like war or death to insipid material like an ad for a hemorrhoid remedy or toilet paper.[56]

Overwhelmed, Overstimulated, and Multitasking

There are many sources of sensory stimulation. Shopping in a Macy's or a Target or a Lord & Taylor retail store tends to make me fatigued quickly because of the bright lights, glittering abundance of products, and ambient noise. Shelves and rows and aisles of really neat, shiny, attractive products compete for my attention. Such abundance overloads my senses and I get tired and grouchy. My wife has a different reaction. She is like many other shoppers who get uplifted and excited by all the stimulating signals: Buy me! Look at me! This is really neat! Many people go to the mall for diversion, entertainment, and recreation. And I do too, but I succeed by keeping a tunnel vision, my focus on that pair of black socks that I set out to buy. I screen out bargains, the video monitors near cash registers, the background music, the lights and signs, all of which compete for my attention like the mermaids calling out to Odysseus.

Increasingly our personal worlds and work worlds are becoming like the stimulating noisy malls—all sorts of things clamoring for our attention—phone calls, unanswered e-mail, new movies and programs, and so forth. The current explosion of all these digital technologies and sources of stimuli not only is changing the way we live and communicate but also is profoundly altering our brains and fragmenting our attention.[57]

It is as if the culture is training our children for digital attention deficit. As our attention fragments, pulled by many stimulating messages and attractions, we try to multitask, that is, do several things simultaneously. Researchers confirm how much digital exposure adults and children regularly receive. "Now we are exposing our brains to technology for extensive periods every day, even at very young ages. A 2007 University of Texas at Austin study of more than 1,000 children found that on a typical day, 75 percent of children watch TV, where as 32 percent of them watch videos or DVDs with a total daily exposure averaging one hour and 20 minutes."[58] Among those children, five- and six-year olds spend an additional 50 minutes in front of the computer. Other studies have found that young people 8–18 years of age expose their brains to eight and a half hours of digital and video sensory stimulation a day.

This flood of visual and aural information plunges us into a state of continuous partial attention which means an effort to keep tabs on everything while never really focusing on anything.[59] The term *continuous partial attention* was coined by Linda Stone in 1998[60] and refers to a cognitive skimming the surface of data for something that seems relevant or interesting, and then moving on. This is different from multitasking where we have a purpose for each task and are tying to improve our efficiency and productivity. Instead, with this other approach, our minds partially attend, and do so continuously, as we scan for any type of contact at every given moment. Many people try to keep tabs on so much through peripheral or fragmented attention.[61]

But multitasking doesn't work very well. When we think we are doing two things at once, we are just switching rapidly between tasks, while leaking a little mental efficiency with every switch. Our brains process different kinds of information by means of separate channels, as it were, a language channel or a visual channel, and an overburdened channel becomes inefficient and mistake-prone.[62] Driving and using a cell phone has the effect of crossing channels, because steering and dialing are manual tasks and looking out the windshield and reading the screen are visual tasks. Multitasking can work if the simple tasks operate on entirely separate channels, such as folding laundry (a visual manual task) while listening to radio news (a verbal task).[63]

Conclusion: Fragmented Life Organized around Different Screens

Just as the preeminence of the feature film as a delivery system for complex visual narratives has been eroded by television, so iPods, smart phones, and websites like YouTube and BitTorrent.com seem to be pushing television aside and changing the experience of moviegoing, according to some who fear the decline of traditional ways of watching films in a theater.[64] A coherent national audience is now fragmented among hundreds of television channels, and multiple new sources for media continue to appear—including the iPad and electronic devices for e-books. Hollywood is shrinking films and television shows for the small screens of cell phones. Vivi Ziegler, a vice-president at NBC Universal, says, "We thought they'd be watching video clips in the checkout line or between classes,"[65] but owners of iPhones and other hand-held devices are watching long episodes and sometimes complete films.

Cultural critics speculate whether traditional media have become irrelevant. Does the content of classic 20th-century media platforms such as Hollywood large screen movies and glossy magazines easily translate onto their digital platforms? Virginia Heffernan, the *New York Times* media critic, says, "It's not possible to translate or extend traditional analog content like news reports and soap operas into pixels without fundamentally changing them. So we have to invent new forms."[66] Thrillers don't seem to flourish on Hulu.Com and no one seems to be reading a six-part investigative series on Twitter. The new digital platforms that enable one to watch a movie on a handheld viewer will probably prompt new content to emerge that is in sync with the new ways of experiencing it. We are probably experiencing the extinction of some ways of organizing words and images, such as "music reviews, fashion spreads, page-one news reports, action movies, late-night talk shows . . . designed for a world that no longer exists. They fail to address existing desires, while conscientiously responding to desires people no longer have."[67] Change happens as new modes or channels of distribution emerge.

As I reflect on the media-rich environment in which we live, two metaphors come to mind for modern *stimulating* life: a lively shopping mall that offers so much that clamors for our attention or a noisy cocktail party where we are trying to focus our attention on a single talker among a mixture of conversations and background noises. The term *cocktail party effect* describes our ability to focus amidst this swarm of stimuli.[68] For example, when we are talking with our friend at a crowded party, we can still listen and understand what our friend says even if the place is very noisy, and can simultaneously ignore what another nearby person is saying. Then if someone calls out our name from the other end of the room, we also notice that sound and respond to it immediately. But we often miss what our friend is saying and we can get distracted by tidbits of a conversation next to us.

How do we deal with altering patterns of attention? How do we keep our focus amidst the new media literacy and changing norms?

We know from brain studies with those who have attention-deficit disorders that excessive stimulation impairs the working memory. People with attention-deficit hyperactivity disorder or ADHD can't hold groups of sentences and images in their minds long enough to extract organized thoughts. Indeed, people with ADHD are often attracted to activities like television and computer games that do not tax the working memory.[69]

This prompts a broader social question: With so much information washing over us and so much stimulating data available, one wonders about the ability of our brains to form certain kinds of significant memories. Perhaps this plays a role in the lack of interest in history on the part of many Americans. Current phenomena have more impact than key aspects of the past: for instance, twice as many people know Paula Abdul was a judge on American Idol than know that the phrase "government of the people, by the people, for the people" comes from Lincoln's Gettysburg Address. This tidbit is from a study by the Intercollegiate Studies Institute. One of the members of that institute, Josiah Bunting, laments that "There is an epidemic of . . . historical ignorance in our country."[70]

An earlier world with little change was probably boring because of a lack of stimulation. Now we do not have *that* problem. "The changes have accelerated dramatically over the past decade, when photographs from planes and satellites have merged with the potential unleashed by faster computers, widespread broadband, and 3-D digital graphics to get us closer to what Google Earth product manager Peter Birch describes as the ultimate aim: an ability to 'recreate the real world in a virtual world.'"[71]

We enjoy and are interested in the new, and thus we are entertained and listen to the news—and many other stimulating ideas and pictures and sounds. But the sheer volume and intensity of the new can overwhelm and change us.

4

Work: How It Changes and How It Changes Us

Peering at a Computer Screen

Technological changes have changed me and my work as a college professor. And change has touched not just me, but the whole American workplace. Before I began teaching more than three decades ago, higher education was a simpler profession, and the rare computer was a giant room-sized mainframe that did not yet dominate the work life of so many.

New technologies and social forces have transformed work. These changes at work are intensely personal as well as institutional.

During my regular work day, when I am not home writing, I now sit facing a computer screen, on average, about three hours a day. I answer e-mail, consult my college's electronic bulletin board, labor at some academic reports, prepare PowerPoint slides for classes, and post assignments on WebCt (a digital site to which students go to retrieve articles, find their grades, and make comments in electronic discussion groups). After teaching several classes I return to my office for further advising, reading, correcting, and the computer screen.

My college, like countless other workplaces, keeps transforming itself in response to forces of change, with new requirements for hiring staff, new procedures for admissions and submitting grades, and with new students, new courses, new leadership. To be hired now is far more complex than when I joined the faculty: for instance, criminal background check, survey of Google or Facebook for indiscretions, a photo ID for security purposes, an orientation to the computer system—all are now required. Increasingly it is routine for companies to set up electronic sites for candidates to submit their digital resumes. At my college, for recent opened new positions, a website was created that listed the prerequisites of the position (including earned PhD, teaching experience, and three references). Candidates sent in their materials electronically; these materials were screened so that the search committee could receive the names of a manageable number of qualified candidates.

The Very Nature of Work Is Changing

Change in work is unremitting, as new technologies and global competition compel industries and individuals to adapt. A long time ago fathers handed down their skills and tools to sons in a world of physical labor. That pattern changed with the last industrial revolution of the 1880s when large companies increasingly divided up manufacturing tasks into specialized jobs. No longer did a single craftsman construct the whole product, be it baby carriage or car or kitchen appliance. Farmers became more productive, and excess laborers—women as well as men—moved to the cities to work in factories. After World War II, the U.S. economy began to accelerate the trend of moving toward service jobs, away from manufacturing and farming, although the majority of people in the world continue to earn their living by their muscles rather than their heads, by physical labor rather than by managing information, as in first-world countries.

Information and the Service Economy

A computer now sits on virtually every desk. The creation, management, and processing of information plays an increasingly large role in our service economy (about 83% of American workers are in the service sector), while manufacturing and agriculture occupy an increasingly smaller proportion of the American work force.[1] Retail and sales workers need upgraded skills and increasingly use sophisticated technology that tracks consumers' purchases and preferences. At Home Depot a cashier will scan the product code on a piece of lumber with an electronic wand that transmits relevant data to the computer, which registers the price. When a customer at MacDonald's requests a Big Mac the computer sends a signal to food-handlers as well as computing the amount the customer is charged. Reward cards, which track sales and customer records, are common in chain stores. They also offer select discounts.

The use of computers and hi-tech equipment has spread to work areas that were traditionally slow to change. American farmers, for example, make use of computers and sophisticated GPS technology in planting thousands of acres. Handheld scanners are used by FedEx and United Parcel Service workers. Police cars regularly have a laptop bolted to the dashboard to check registrations and shared databases. The modern auto repair shop increasingly relies on a computer for customer records or for diagnostics on contemporary automobiles' numerous chips and minicomputers that regulate engine performance. No area of contemporary work is untouched by electronic and digital technology.

Almost 70 percent of the U.S. labor force are now considered information workers, that is, people who work at tasks of developing or using knowledge. For example, an information or knowledge worker is someone who works at any of the tasks involved in "planning, acquiring, searching, analyzing, organizing, storing, programming, distributing, marketing, or otherwise contributing to the transformation and commerce of information."[2] A term first used by Peter Drucker in his 1959 book, *Landmarks of Tomorrow*,[3] the *knowledge* or *information worker* specifically includes those in information technology, such as programmers, systems analysts, technical writers, and so forth. The term *information worker* also commonly includes people not specifically in information technology, such as lawyers, teachers, scientists of all kinds, and students of all kinds.

The U.S. Government's Bureau of Labor Statistics provides tables of occupations that paint a portrait of Americans at their jobs.[4] Occupations with the fewest number of workers include watch repairers, astronomers, and radio operators. In 2009 the occupations with the highest employment included workers whom virtually everyone of us comes into contact with regularly: retail salespersons, cashiers, general office clerks, food preparers and servers, and customer service representatives. Most of these largest occupations were relatively low paying, with average wages well below the U.S. mean of $20.90 per hour or $43,460 annually. Generally those who earn above-average wages are those who deal directly with change, for instance, general managers and operations managers. Those workers who hover near the bottom of the pay scale are cashiers who earn an hourly average wage of $9.15, and combined food-prep and serving workers who earn $8.71. But many of these workers at the low end of the earning scale regularly use some electronic device that facilitates efficiency, such as the MacDonald's worker who wears a headset and lists orders from the takeout window on a computer screen. Restaurant waiters commonly use handheld gadgets that communicated orders directly to the kitchen. This use of technology has also changed the way service calls are made in customers' homes. Even as I write this, there is a furnace repairman in our basement installing a small computer to make our furnace more efficient; he is connected to his manager and home office by a Pegasus computer system that relays data and billing information.

Noticeable in labor statistics is the increase in those workers whose work directly depends on computers and information technologies, even in manufacturing: thus, in 2009 there were 129,780 computer-controlled machine tool operators for metal and plastic fabrication. In manufacturing there is a long-term movement toward greater automation, with the laying off of the lowest skilled workers—and even replacing these workers with cheaper foreign labor. Manufacturers seek workers who can operate sophisticated computerized machinery and follow complex blueprints. Higher-skilled employees replace the traditional assembly line worker.

Economists have sought to assess the productivity of the economy as it shifted toward service and toward processing information by the use of computers. That shift presented economists with a conundrum in an era when there appeared to be a contradiction between the remarkable advances in computers and a relatively slow growth of productivity. The concept is sometimes referred to as the Solow Computer Paradox in reference to economist Robert Solow's 1987 quip, "You can see the computer age everywhere but in the productivity statistics."[5]

Economists, however, soon figured out that productivity gains were indeed real and that many of the advances were slowly working their way through the economy. In brief, there is some lag between the introduction of a new technology and its visible implications. Using data-mining techniques to analyze patterns of electronic information flows within firms and between firms, economists soon realized the large impact that computers and the Web had made on the economy.[6]

Generally speaking, workers at the lower end of the economic-education scale have the least security and are easily displaced by more efficient technologies. New technologies have unexpected consequences as they become widespread and evermore sophisticated. Computers and digital devices like BlackBerrys increased productivity, but often at the cost of uncertainty and stress as companies need fewer employees—but more employees with complex skills. Many of the jobs lost in the last recession are not coming back. The weakened economy provided the opportunity for employers to dismiss millions of workers like file clerks, ticket agents, autoworkers who were displaced by technological advances, and international competition. Increasingly the contemporary office is more automated and digitized than ever. Journalist Catherine Rampell remarks, "Bosses can handle their own calendars, travel arrangements and files through their own computers and ubiquitous BlackBerrys. In many offices, voice mail systems and doorbells—not receptionists—greet callers and visitors."[7]

When there is transformation or turbulence in the marketplace, there will be displaced workers who work as hard as they can but feel that events are racing beyond their control. How many American workers have felt this in recent years—say, in the steel or manufacturing industries? Autoworkers who make gas-guzzlers that do not sell know the union cannot protect them from potential plant closings. Cable television, Video-on-Demand, and NetFlix have all pushed video rental stores to the brink of extinction. "Casual Friday" means fewer tailors are needed to manufacture men's suits and formal wear.

Because so many products Americans use and so many of the very clothes Americans wear are manufactured overseas, the hot breath of global competition can be felt in many of the local shops and factories that used to make these items.

Many New and Specialized Jobs in a Complex Society

The Bureau of Labor Statistics conducts research on what new occupations are emerging and which should be included in the government publication

Standard Occupation Classification.[8] This publication uses federal research and statistics to classify workers into 23 major groups of occupations, with sublistings for 840 occupations. What forces drive creation of new and emerging occupations? What are examples of these new positions and their requirements? How does the Bureau of Labor Statistics determine types of new or significantly changed occupations? For example, a new occupation includes duties that have developed recently and are not included in a current occupational classification. Retinal angiography, a subfield of biomedical photography, is an example. Some occupations that are already recognized but have continued growth can be referred to as emerging occupations. Thus "search engine optimization" is established enough to have its own professional association and trade group.

Young college graduates move into an ever more complex world of work, with specialties simply not foreseen just a few years ago. Technological advances, new laws and regulations, and changing demographics tend to give birth to new occupations. Occupational areas with significant revisions and additions include, as might be expected, information technology (especially computer occupations) and health care (especially technical and support occupations). Events that contribute to new employment opportunities include natural disasters, war and terrorism, and global competition. New and emerging occupations are increasingly multidisciplinary, specialized, and international. Other new occupations listed in 2010 *Standard Occupation Guide* are Web developers, computer network architects, exercise physiologists, magnetic resonance imaging technologists, hearing aid specialists, transportation security screeners, solar photovoltaic installers, and wind turbine service technicians. In other words, this guide simply codifies what people are actually doing and have been doing for some time—but not officially named and categorized until now.

Number of Jobs Held in a Lifetime

It appears that people are changing jobs more frequently. To determine the number of jobs in an average lifetime, one would need data from a longitudinal survey that tracks the same respondents over their entire working lives, and so far, no longitudinal survey has ever tracked respondents for that long. A Bureau of Labor Statistics news release published in

June 2008 examined the number of jobs that people born in the years 1957 to 1964 held from age 18 to age 42.[9] These younger baby boomers held an average of 10.8 jobs from ages 18 to 42. (In this report, a job is defined as an uninterrupted period of work with a particular employer.) On average, men held 10.7 jobs and women held 10.3 jobs. Both men and women held more jobs on average in their late teens and early twenties than they held in their mid-thirties. Almost a quarter held 15 jobs or more. By contrast, in Asia people tend to stay with a company for a longer time. They seem to have a stronger sense of loyalty to their workplaces. In Japan, workers have this "I sold my life to the corporation" mentality, and many stay with the same corporation from when they first started working right after graduation from college till the day they retire, climbing the ladder within the same corporation. Changing one's corporation or the place to work is like changing the world for them, and the employment situation is not as flexible as in the United States. If one's work is an aspect of identity, what meaning does this shifting of jobs have to one's sense of self and continuity? I shall return to this question in the next few pages, but it makes me think of my own decades-long service at one college; by contrast, both of my sons have had many more job changes than I have had.

Who Manages Change?

One group in the world of work that is especially sensitive to change includes managers or business executives. Each year numerous books and articles address those who lead companies that, amidst competition, must constantly jockey for profit and survival. "It is an accepted tenet of modern life that change is constant, of greater magnitude and far less predicable than ever before. For this reason, managing change is acknowledged as being one of the most important and difficult issues facing organizations today," remarks author Patrick Dawson.[10]

Research shows that organizations change primarily in two ways: through drastic action and through evolutionary adaptation. In the case of drastic action, "change is discontinuous and often forced on the organization or mandated by top management in the wake of major technological innovations, by a scarcity or abundance of critical resources, or by sudden changes in the regulatory, legal, competitive, or political landscape.

Under such circumstances, change may happen quickly, and often involves significant pain. Evolutionary change, by contrast, is gentle, incremental, decentralized, and over time produces a broad and lasting shift with less upheaval."[11] No matter how much an institution attempts to shield its employees from the uncertainties and stresses of change, there is still anxiety. Executives and employees tend to see change differently, with senior managers thinking of change as opportunity, while employees may see change as disruptive, intrusive, and threatening.[12]

Failure to evolve and adapt to a changing world can lead to extinction, just as with the dinosaurs. Of the 500 largest companies listed in *Fortune Magazine* in 1955, only 71 are still listed today. Many behemoths fell from the list, such as Zenith, Chrysler, and Bethlehem Steel. And growing from zero to greatness upon entirely new technology are Intel, Microsoft, Apple, Dell, and Google.[13] The landscape keeps changing as fabled and faded names in U.S. business (Montgomery Ward, Compaq, Sun Microsystems, America Online) get gobbled up by larger, hungry competitors.

The technology that triggered the creation of many new business upstarts was the integrated circuit or silicon chip, invented by Robert Noyce and Jack Kirby in 1958.[14] Integrated circuits built upon previous discoveries that advanced the functions of vacuum tubes and allowed for the combining of transistors, resistors, and capacitors on a single silicon wafer. This integration paved the way for the first microprocessors in the early 1970s, which in turn gave birth to smaller and faster computers—which in turn unleashed a gale of creative destruction. *Creative destruction* is a term used by Joseph A. Schumpeter[15] to describe the kind of revolution that occurs in the process of innovation in industries and businesses. Competition, according to Schumpeter, comes from "the new commodity, the new technology, the new source of supply, the new type of organization . . . which commands a decisive cost or quality advantage and which strikes not at the margins of the processes and the outputs of the existing firms but at their foundations and their very lives."[16] Thus, Xerox Corporation held an advantage in copiers but missed the competitive boat when rivals produced smaller and cheaper copiers. Polaroid held absolute sway in instant photography until other nimble competitors produced better designs and the digital camera. Or, a favorite example of mine is in music, where the long-playing record replaced 33 RPM and 45 RPM records, only to be replaced in turn by the compact disc, which in turn was undercut by MP3 players. Most of the time, these advances are better for consumers. But the changes and competition often lead to uncertainty for both consumers and workers, and for some companies, bankruptcy.

The Changing Work Environment from a Consumer's Viewpoint

Changes in the workplace give birth to different experiences for consumers. First, let's consider food shopping. My wife and I have lived in our neighborhood for 30 years—a long time. Our local supermarket has undergone several takeovers by large corporations, renovations, and name changes. I have shopped at this store for decades, because it is convenient and I like the people who work there. After changes of ownership and procedures, I chatted with the employees, who were distressed because they had no local control and had to deal with complaints from customers who didn't like some of the changes initiated by managers in a distant city. This store again is enduring a renovation—at least the third one since I have been patronizing it. I zigzag down aisles that have been revamped and supposedly improved. I fully acknowledge how nice and burnished things look, but it is taking me longer than I expected to master the new layout. Of course, I will carry on, but I also notice that a number of products I regularly buy are no longer available. I asked the manager about this. After he checked the computer, he sheepishly came back and said, "Oh yes, well, there is only so much shelf space and we will not be carrying that product any longer. But there are some others just as good . . ." Many consumers notice how, because of shifting market forces, some favorite products have disappeared, never to be seen again; I recognize these small inconveniences as the price of living amidst such abundance and choices.

Among other changes I notice while walking around my renovated local supermarket are women now working in jobs that used to be done by men, for example as construction workers, as butchers, or as managers.

Women now make up half of the U.S. labor force and are projected to account for most of the increases in the labor force over the next few years. Forty percent of employed women work in management, a category that includes professional and related occupations, while the largest single occupation category is composed of secretaries and administrative assistants, with more than 3 million women.[17]

I love bookstores, but several of my favorite bookstores have vanished in the last few years. One of the best, however, is still open about nine miles away. What I like about this store is the opportunity to browse and see new books that I had no idea were available. But I have many more choices from Amazon.com, the online seller of books and other products. No retail store can have the range and number of both new and used books as Amazon—but a customer must have some idea of what to seek amidst such an abundance of possible choices. There used to be 4,700 independent bookstores in 1993; now there are 2,500.[18] Between 1991 and 1996 the independent booksellers lost 42 percent of the market share as the Internet burst onto the economic stage in the form of the new "e-economy," which is expected to increase 1,000 percent. The most successful companies have to react quickly to embrace these new realities.

Thus retail stores now scramble to be entertaining destinations in order to regain customers who almost certainly have other options, such as Internet sites or catalog stores. Jacques Barzun notes a contemporary blurring of distinctions. He cites a tendency toward what he calls "conglomerate experience," that is, the pattern of mixing pleasures, activities, and items-for-purchase, all available in one place. Thus, there used to be a general store or department store with a mixture of goods, but no other amenities; now art museums have added sales of books and posters *and* offer cafeteria food; libraries offer more than books when they have coffee bars; universities provide an education *and* offer their alumni guided tours of regions of the world; bookstores provide not only books but also toys for children and coffee for adults.[19]

Also from my perspective as a consumer, I am surprised when I chance upon a product manufactured in America. Globalization, changing labor markets, and industrial restructuring have impacted the American worker. Indeed, American manufacturing has declined, employing scarcely 10 percent of American workers, down from 24 percent of workers 40 years ago.[20] Although the United States still is a leading manufacturer, especially of high-end products like Boeing jet airliners or Caterpillar earth movers, retail shelves are stacked with relatively inexpensive clothing and everyday products from abroad. When I go shopping with my wife I tend to look at labels to see where products are made. "Made in Japan" used to be the common label, but now "Made in China" is what I find about 90 percent of the time. Nearly 90 percent of laptops and digital notebooks, made under famous brand names, are manufactured by one of five companies in China. So far this is still good for U.S. consumers. James Fallows tells of an auto device that in the United States sells for $30, of which $6 goes to the Chinese manufacturer, while $24 stays with the U.S. importer. High-end Ethernet connection cables sell for $29.95 in the United States, but the Shenzhen Company of China only makes $2 from this sale. Low-paid Chinese workers help American designers, marketers, and retailers.[21]

Some of the low prices may be related to the average life of a consumer electronics product, which is less than nine months. Decades ago Ford's Model T automobile was unchanged for almost 20 years, whereas at Hewlett-Packard today nearly two-thirds of orders are for products introduced in the last two years. Apple introduces new upgrades or new products every few months. "What has changed about change is its rate of introduction, and the ripple effect on so many aspects of our lives. . . . We feel the impact of this changing nature of change every day . . . of our work lives."[22] This keeps workers and companies scrambling to invent, design, and deliver new products to stay competitive and keep prices low.

Payment and Electronic Transfers

More convenience and enhanced modernity results in churn and gyrations in the stuff of everyday life. This changing scenario offers new and shifting opportunities for workers—but also uncertainty. A good instance of this twisting and shifting is how we pay for products and services. We have the convenience of ATM machines, electronic transfers of funds, direct deposit of paychecks, and online banking. Online banking has enabled consumers to check accounts and manage tasks that formerly required a bank employee; online banking also helps create both new companies that seek to supply and service ATMs, and new careers such as programmers for electronic services.[23] One senior bank officer told me that she notices many fewer customers now come into the bank to do their business. "The young ones do it all online and the old ones are dying and disappearing."[24]

When I went to see my physician recently I noticed the patient ahead of me paid with a credit card, another paid by check, and one person paid by cash—which is increasingly rare. When I saw my physician she was muttering about new procedures that insurance providers were demanding before she herself was paid. She told me that when she began the practice of medicine she could see a lot of patients, but now regulations were such that she had to do extensive record keeping for proper payment. Further, even though she was conscientious about keeping up with the continuing medical education requirements, to retain her medical license she had needed to be recertified because some insurance providers would not pay physicians who had not been recertified. Medicine, like other professions, has changed and continues to change in many ways, including payment procedures.

Credit cards and electronic payments are replacing cash transactions; one can even pay for a cup of coffee with a debit or credit card. In 1949 a New York businessman, Frank McNamara, went to pay the bill after entertaining a client only to realize he did not have his wallet with him. He had the idea of a club of diners who would be able to sign for their meals at certain restaurants and then settle the bill at a later date. As his idea became a reality, McNamara enrolled 27 establishments in his plan, offering $3 memberships in his diner's club to 200 friends and acquaintances. "Charge it," said McNamara and his friends when they sat down to a February 1950 meal at a local restaurant—the first diners of many millions to say those magic words. With 20,000 cardholders by the end of 1950, the Diners Club was an instant success. On the heels of Diners Club, American Express introduced a card in 1958 for paying entertainment and travel costs. The next year, Bank of America issued a revolving credit card that could be used for a greater range of purchases and paid off over a longer period of time, with interest. Credit cards became more common in the middle 1960s.[25] The rise in telephone-connected computers made electronic transfers of cash easy. Now there are many, many options for payment, from debit cards to the use of different kinds of rewards cards and loyalty programs.

"The major card networks have been pushing to use plastic for everything," said Gwenn Bezard, director of research at Aite Group LLC, a Boston consulting firm. Cash accounted for 35 percent of the 137 billion payment transactions in the United States in 2006, down from 44 percent of 117 billion transactions in 2001. Meanwhile, payment methods such as those involving credit, debit, and prepaid cards rose from 29 percent to 42 percent of all transactions during the same period. The number of ATMs reflects those trends. After peaking at 396,000 in 2005, the number of ATMs –around for 30 years—began to decline, because of the increased use of debit cards and online payments.[26]

But the real revolution in retail, other than the omnipresence of computers, global competition, and decrease in the use of cash, is in retail purchases via the Internet. Granted, most food shopping is not done on the Internet, yet brick-and-mortar stores have had to scramble to become entertainment destinations to attract customers who increasingly switch to Internet and catalogue sales.[27]

Measuring the Electronic Economy

When people pay for something via the Internet, they are joining the electronic economy or *e-commerce.* The U.S. government census website defines e-commerce as "sales of goods and services where an order is placed by the buyer or price and terms of sale are negotiated over an Internet, extranet, Electronic Data Interchange (EDI) network, electronic mail, or other online system."[28] The *E-Stats* website is the U.S. Census Bureau's Internet site devoted exclusively to "Measuring the Electronic Economy." The site[29] reports that in 2008, e-commerce grew faster than total economic activity in three of the four major economic sectors covered by the *E-Stats* report. U.S. retail e-commerce sales reached almost $142 billion in 2008, up from $137 billion in 2007, and increased again in 2009 to about $151 billion. From 2002 to 2008, retail e-sales increased at an average annual growth rate of 21.0 percent, compared with 4.0 percent for total retail sales. In 2009, e-sales were 4 percent of total retail sales. The number one online transaction category was travel, with 38 percent of adult online consumers making at least one travel purchase on the Web in the previous six months. Credit card account management and home banking took the No. 2 and 3 spots, with 36 percent and 35 percent of consumers conducting transactions, respectively. Nearly 80 percent of U.S. adult online consumers made Internet purchases in the pervious six months, according to Nielsen Online, a service of the Nielsen Company.[30]

Online sales, which slowed in the economy's downturn, have still managed to outpace sales at U.S. brick-and-mortar stores. These numbers are startling: e-commerce is definitely here to stay, and the global marketplace is now at one's fingertips. As security measures improve and online shopping technology improves, an increasing part of U.S. economic activity will be online—an extraordinary transition in a very short period of time.

Workplace Environments: Multitasking and Fragmentation

The modern workplace is saturated with multitasking and information technologies as workers try to keep aware of information and maintain the nearly instant communication that supports collaborative work and productivity. Perhaps there are noisy conversations, or switching among multiple windows on a computer, or someone may need to attend to a message-alert on a cell phone or computer window. And looming in the background is the expectation of immediate responses to e-mail and messages. Multitasking, distractions, and interruptions of tasks at hand have become routine and commonplace. Consequently, researchers scramble to find out just how often interruptions are, how long they are, what the effect is on attention, and what the recovery process is after interruptions.

Researchers at the University of California's Donald Bren School of Information and Computer Sciences are now investigating the environment of information workers. They discovered that information workers—and all types of workers, in fact—experience a lot of discontinuity in their work: "Information work is fragmented work," comment Victor M. González and Gloria Mark.[31] People average about three minutes on a task and somewhat more than two minutes using any electronic tool or paper document before switching tasks. The researchers use the idea of a *working sphere* to explain how individuals organize the basic units of their work. During a single day the people they studied worked in an average of 10 spheres. These spheres, however, were also fragmented. People spent about 12 minutes in a sphere before they switched to another. In brief, many, many interruptions occur in information work of any kind.

The California researchers highlight the cost of interruptions and the efforts of people to switch among tasks, that is, multitasking. In the "attention economy," information work is fragmented work. Fragmentation is a break in continuous work activity, and information workers are especially prone to spending short amounts of time on tasks with frequent switching. Generally, depending on the discipline, information-work environments are fast-paced and characterized by multiple conversations, telephones ringing constantly, and people walking into other cubicles unannounced. At one particular worksite, the high money value of information played a role in shaping the frantic rhythm of work. In areas of financial analysis and high-value trades, the work had to be accurate and done quickly.[32]

Other areas of work are less media-oriented. To be sure, multitasking is common and we all do several things at the same time: driving while listening to a radio, talking on the phone while watching a child, watching television and talking to a spouse. Modern life, with multiple providers of information, offers increased situations where we attempt to multitask. The work situation, with modern information technologies and ways to connect to fellow workers, puts us in situations that tax our brain's response to many of the demands placed on it. Many of the interruptions involve more information: overheard conversations, ringing phones, blinking alerts to new e-mail messages, our iPods playing music and multiscreens opened on our computers. . But how many tasks and what kinds of tasks can we do simultaneously and be effective?

Short-order cooks must keep track of a half-dozen orders in their brains as they crack eggs and toast muffins and refill the coffee maker. But these cooks simply do what a fast computer does: switch between tasks. It is not possible to do more than one principal task concurrently. Actually we do rather well at shifting focus from one thing to the next with astonishing speed. Switching between tasks that are of a different mental level is more manageable because that is primarily a physical task not involving complex thought, as contrasted with a mental task demanding close attention. Beginning short-order cooks need all the brain participation they can muster, as they learn their job. But as they gain experience, the level of focus changes. The brain rewires itself to do the tasks involved—especially if it is a routine physical operation—and many manual tasks become automatic. This is true, say, for the novice automobile drivers who need to turn off the radio, not talk, and be very attentive to the accelerator and the brake and steering. Over time, increasing aspects of driving become automatic. The mind gradually allows us to do one thing automatically while focusing on something else—for instance, arriving at work and not remembering much about the trip.[33]

What about information workers who experience multiple activities and interruptions in their work? What is happening to the mental processes involved in interrupted mental work and multitasking? Are people deluding themselves on how well they can multitask? Any interruption introduces a change in work patterns and mental concentration. Those interruptions that change the cognitive processes of focusing and attention seem to be more disruptive. When the content of an interruption and the task at hand were consistent or similar, interruptions could be beneficial. Thus interruptions that share a context with the main task might be seen as beneficial, but the overall disruption in efficiency and energy is costly nevertheless.

Surprisingly, sometimes interrupted work is performed faster.[34] When people are constantly interrupted, they develop a mode of working faster (and shrinking the amount of writing on a project) to compensate for the time they know they will lose by being interrupted. Yet working faster with interruptions has its penalties: researchers found that people who were interrupted experienced higher workload, more stress, higher frustration, more time pressure, and greater effort to finish their tasks. Therefore interrupted work may be done faster, but at a price. Some interruptions lead people to change not only their work rhythms but also strategies and mental states. Interruptions often, in fact, lengthen the time to perform a task, but this extra time usually occurs after the interruption when the worker turns back to the original task, and then the interruption can be compensated for by a faster and more stressful working style.

Thus context determines whether interruptions and multitasking may be beneficial or detrimental. Task switching may be beneficial, if it serves to refresh the worker and provide new ideas. On the other hand, too much task switching with too many activities could be detrimental. Task switching that requires start-up time in a complex project could result in a lower level of accomplishment. Such interruptions can lead to the stress of keeping track of multiple stages of tasks. Especially disruptive are those distractions and interruptions during tasks that require a lot of concentration—such as solving a production problem. The interruptions that result in losing one's train of thought for the task-at-hand are costly, especially if the interruption occurs when it is not a natural breaking point for a task. The most disruptive of distractions are those that lead the worker to shift working spheres or mental set. The cost comes in difficulties remembering what the workers were doing in their interrupted task. Some of the workers described to researchers that they began working on yet another task until they remembered what they were originally doing. Sometimes this memory lapse led to redundant work. One engineer said that he sometimes might "forget what I was testing and I might retest the same thing." Another said that he might have his mind on something else, "and then you have to shift completely. It is disruptive in the sense that if we are going to leave it unattended for a period of time and by the time you come back to it your frame of mind is completely different." A third worker offered: "You forget what you are working on so you kind of do something else for a while and then you remember what you were working on."[35]

Researchers say that frequent interruptions and multitasking cause a decline in performance that is "the equivalent to losing ten IQ points—two-and-a-half times the drop seen after smoking pot."[36] Concentration on a mental task, such as learning complex material or seeking to solve a problem, is not easy. Many of us will flee—temporarily—the effort required by such tasks. Sherry Turkle, the MIT researcher, in an interview on PBS, describes her efforts to make a difficult or demanding point in her writing: she leaves off her writing to go to her e-mail and, after 20 minutes further losing her train of thought, she decides to check Amazon. . . . "It is a universal seduction," she says, "to do every little thing to break up the difficult mental tasks."[37] Merely getting up to stretch or get coffee allows one to stay with the problem or clear one's mind. But shifting to e-mail or browsing Amazon causes one to lose focus on the problem for a time. Turkle believes we're getting ourselves out of the habit of just staying with something hard. One wonders about the quality of homework accomplished by students who have cell phones at hand, their iPods playing music, and a television flickering in the corner of the room. Or the students who have several open windows on their laptops, with another window upon for iTunes, and perhaps a game of solitaire. Or is the screen open to Facebook, YouTube, and instant messaging for comments to a friend?

Some studies suggest the human mind is uniquely set up to do two things at once—but definitely not three, at least not if these include mental tasks that require concentration. French researchers, through the use of imaging techniques on volunteers doing just one task, discerned activity in goal-oriented areas of both frontal lobes of the brain.[38] This suggested that the two sides of the brain were working together to get the job done. Because the brain has only two frontal lobes, might there be a limit to the number of goals the brain can handle? The French researchers added a second goal. When the volunteers pursued two goals, the images showed the brain assigning oversight of one task to the left frontal lobe and the other to the right frontal lobe. This division of labor suggests that humans would struggle when asked to carry out more than two tasks at one time. The researchers confirmed this when the volunteers attempted a third simultaneous task. Images that had indicated activity in the brain on one of the original goals disappeared, reflecting the brain's inability to work on three tasks at once. Furthermore, the volunteers slowed down and made many more mistakes.[39]

Such research does not show subjects able to do even two tasks simultaneously. The brain can only maintain two goals at once: the mind switches back and forth between the tasks themselves. Researchers routinely find there are large costs involved in trying to do this switching between two tasks, making multitasking inefficient. Some commonplace activities, such as driving and talking on a cell phone, frequently go hand-in-hand, but the brain is likely to be switching its main focus quickly between the two activities, rendering this particular pairing dangerous.[40] Driving a car requires a large amount of brain power. We need to process huge amounts of visual information, predict the actions of other drivers, and coordinate precise movements of our hands and feet. "If you're driving while cell-phoning, then your performance is going to be as poor as if you were legally drunk," says David Meyer, a psychology professor at the University of Michigan.[41]

The issue comes down to the complexity and nature of the tasks that the brain is attempting to accomplish. The brain's frontal cortex exerts executive control like a control tower, helping the person to stay focused and to set priorities. Survival instincts, however, which for eons enabled humans to stay alive, respond irresistibly to movement and sights and sounds.

Evidence suggests that human cognition is ill-suited both for attending to multiple sensory streams and simultaneously performing multiple tasks.[42] Modern technologies provide information channels that supply streams of images and data that easily overwhelm our brain's capacity. Our brains evolved enabling us to run across the plains tracking prey, avoiding obstacles, and keeping human beings focused on lunch-on-the-hoof. Our brains did *not* evolve to manage multiple streams of information that are unrelated, such as chatting with a number of different people while working on a paper as we watch TV.[43]

Heavy multitaskers have trouble tuning out distractions and switching tasks compared with those who multitask less. And there's evidence that multitasking may weaken cognitive ability. Clifford Nass and other researchers at Stanford University[44] suggest that several cognitive processes are required for focused attention: filtering out irrelevant stimuli in the external environment and irrelevant representations arising in memory. But multitaskers don't filter well and are less effective in suppressing irrelevant task-switching. The bottom line is that multitaskers commonly sacrifice performance on the primary task to let in other sources of information, which leads to inefficiency and error. Thus, those multitaskers who think they are really good at juggling many things at once are probably even slower and worse at switching from one task to another than the rest of us.

Music, or at least instrumental music, seems to be the one modality that doesn't seem to lead to problems with multitasking. So it is fine to listen to music and do intellectual work. But jumping rapidly from playing a computer game to browsing online to writing a Word document to e-mail cuts down on efficiency and leads to potential errors. There may indeed be a whole new mental meaning and significance, then, in those recorded messages that say: "I'm either away from my desk or on another line." Should they more correctly say, "I'm away from my desk *mentally*"?

Blurred Boundaries between Work, Nonwork, and Leisure

Leisure can be considered free time that is at one's disposal. Work (gainful employment) is the category against which leisure is defined. But nonwork activities are not necessarily leisure. So we must add to these two categories a third broad and messy category: necessary activities that have to be done for daily living, such as errands and food shopping, caring for children, and those other tasks for which there is no material or economic return. Nevertheless, for our discussion here, there are anomalous areas in the gap between work and leisure: the growth of people employed in the leisure industry.[45] Professional basketball players or musicians get paid to play and they have a very different experience of a game or gig than the paying audience. (The Bureau of Labor Statistics lists more than 13,000 athletes and sports competitors and more than 80,000 musicians and singers.[46]) Looming even larger as a pesky modern problem are those who keeping working even when they are not at work. Increasingly for some professionals, weekends and vacations mean they can leave the office physically, but not mentally, as they take their modern communication technologies along with them on vacation: here's new meaning to time-on and time-off.[47] In a money economy, however, time and work tend to get linked; time is money, as we shall see in the next chapter.

The nature of work is effort in exchange for payment. Another aspect of work is the sense of being constrained: I'm not doing this task merely for pleasure or satisfaction—although those qualities may be present—but because I am earning my daily bread, caring for my family. Thus, when I am working in my garden on a weekend, it is a leisure activity that I freely take on for its gratification. When I hire someone to do the very same task, the recreational quality of the labor gets replaced by the employer-employee relationship and the different labor quality of the task itself. I can only wonder what our ancestors, who did hard physical labor for pay, would think of the more than 45 million Americans who pay to work out at the nearly 30,000 gym and health clubs in the United States.[48]

Leisure used to be the defining feature of an elite lifestyle, but by the middle of the 20th century this longstanding pattern began to be reversed. The gap between the amount of leisure that the poor had compared to the leisure time of the better-off has grown, with the poor and less educated having more leisure time—whether chosen by them or imposed on them. Although the workweek has stayed constant over the past several decades, Americans get scandalously little vacation time, averaging about 14 days per year. (Germans average 26 days off a year and the French 36 days off).[49] The U.S. Bureau of Labor Statistics suggests that Americans have a comparatively large amount of leisure time, an average of 5 hours and 15 minutes of free time on any given day.[50] But many Americans—especially at the higher end of the pay scale—seem to experience significant time pressure and their work has expanded beyond the workplace. Some employees choose to work from their homes via computers, telecommuting and thus having more control over their time. But the home is invaded by time pressures and procedures of work—while in a strange way the workplace for some becomes a surrogate home.[51] It is common to see people at work in locations that previously have not been worksites: at Starbucks, on the train, at the beach bending over a laptop and earnestly speaking into a cell phone or checking a BlackBerry. The busiest occupational groups tend to be professionals and managers. More than one in three men (37.2%) who work in professional, technical, or managerial occupations put in 50 hours or more per week, compared to one in five in other occupations. "If life seems increasingly fast-paced to the many scholars and observers who write and read about these matters, it is partly because they are members of the group where this experience is quite common."[52] I confirm this from my own experience and that of my colleagues who commonly do research or correct papers at night or on the weekends.

Workers at the high end of the pay scale, especially those in the information- and technology-driven workplace and those with a work-leisure imbalance, can cushion work-related changes by eating out instead of cooking, by hiring housecleaners, and by buying no-wrinkle clothing instead of ironing. The past three decades have witnessed a surge in meals eaten outside home, a growth in home cleaning services, and an increase in prepared, quick meals in supermarkets. The affluent avail themselves of dishwashers and microwaves and other labor-saving appliances to cut down on time spent on domestic tasks. Service industries, like banks, expand hours to include weekends, evenings, and drive-through and online banking.[53]

We Are Our Work—Maybe

Work has changed in the last several decades, continuing the transition from the physical to the realm of knowledge and information, often with a blurring of the distinction between work and leisure, between work and home. The nature and role of work continues to evolve rapidly—especially at the high ends, if the work is interesting and intrinsically rewarding—but for many, work is drudgery. We live in a culture where there are significant disparities in wealth, income, and free time. As companies shed full-time workers for interim or temporary employees—for flexibility and competitive reasons—there is an impact on loyalty, status, and expectations. Perhaps this suits some workers who can tolerate turnover and change. (One young businessman told me that he thought the next generation, used to multitasking and distractions, gets restless and bored more easily and may be quicker to seek out new and more stimulating employment.)

Because work forms a major aspect of who we are, the changes present at our places of work affect us deeply. We tend to identify ourselves by our work: "I'm a computer programmer," we say, or "I'm a teacher." For most of us work plays a central position in our lives, providing meaning and structure to daily life as well as a paycheck. How we spend our time over the course of a week considerably shapes our status, expectations, and identities. If we are somewhat trapped with few options or if we do not have leverage to be proactive in the face of change, work can be draining and stressful. If our work is changing rapidly, it threatens our self-esteem and our sense of meaning and introduces uncertainty.

Increasingly we encounter the global and economic pressures that keep changing the very nature of work. As Ronald R. Sims says:

> As the twenty-first century begins, the world is in a constant state of change, and no organization, in the United States or elsewhere, can escape the effects of operating in a continually dynamic evolving landscape. The forces of change are so great that the future success, indeed the very survival, of thousands of organizations depends on how well they respond to change or, optimally, where they can actually stay ahead of change.[54]

5

New Behaviors and Changes in Manners

Public Behaviors

Change can be seen in the mundane behaviors of everyday life.

I enjoy my commute to work since it gives me a chance to observe people going about daily tasks. On my daily walk to the train, through a middle-class neighborhood, I pass a local elementary school where I notice lines of cars pulling up to drop off children. These children live close enough to walk to school in this safe neighborhood—and there is a bus for those further away. "Forty years ago, half of all students walked or bicycled to school. Today, fewer than 15 percent travel on their own steam. One-quarter take buses, and about 60 percent are transported in private automobiles, usually driven by a parent . . . the change was primarily motivated by parents' safety concerns," says writer Jane Brody.[1] A desire to protect children from traffic hazards and predators is understandable, but behaviors have consequences, such as a proneness to a sedentary style and perhaps obesity. Lenore Skinezy acknowledges that childhood has changed in the last few decades, but regards many parental worries as a product of viewing television, which broadcasts many crime-centered and scary programs—programs that foster parental worries about the safety of their children—even though crime statistics have fallen.[2]

When I get to the train I continue my surveys of fellow commuters riding the Boston T. This train is a good arena to notice emerging behaviors, trends, and fads. It is commonplace now to see half or more riders on the Boston T using cell phones or texting. Increasingly one now sees electronic readers amidst the books or magazines that some commuters scan. Often commuters are so absorbed with their phones that they fall short in the traditional courtesy of offering their seat to a pregnant woman or a handicapped person. Not uncommonly it is an elderly person who is not using a phone who stands and donates a seat to someone needing it.

Another daily transportation phenomenon is public eating and carrying a coffee mug. In the blog-sphere there are debates about the etiquette of eating in public around other people who aren't eating.[3] My impulse is to be tolerant of this form of multitasking as it may be in response to the pressure of all the other things that busy commuters are trying to accomplish. But eating in public raises the twin online discussion of American obesity and fears of anorexia. Michael Pollan[4] laments the American habit of grazing and snacking or watching television during mealtime instead of sitting down for leisurely meals. Pollan argues that food has been more or less invisible until recently, but now the "supermarkets brim with produce summoned from every corner of the globe, a steady stream of novel food products (17,000 new ones each year) . . . and in the freezer case you can find 'home meal replacements' in every conceivable ethnic stripe, demanding nothing more of the eater than opening the package and waiting for the microwave to chirp."[5]

Cell Phone Behaviors

At an off-campus department meeting not long ago, the absence of a senior member of the department troubled me. Belatedly I understood that she counted on a last-minute text message from me to give her directions. Although directions had been sent to her and other faculty weeks ago, she is tied to her computer and smart phone, checking e-mail and messages many times a day, so that she has gotten accustomed to just-in-time decision making. The speed of communication with modern devices makes this kind of behavior possible.

The consequence of instant communication spills over into a tendency toward last-minute decisions. By that term I refer to delayed responses to social invitations because of the use of cell phones and text messaging to stay in immediate touch with friends. I have tried to arrange several social events for undergraduate students in my department and have asked them to RSVP. Very few students respond by any means ahead of time and so I must guess at numbers when I am ordering refreshments. I've asked students why they don't get back to me. Their responses suggest that many of them do not form their social plans until the last minute. They hold off committing to an event or a film or coffee because something else—that is, something better—might come along.

Many in the twenty-something generation exemplify this pattern. The instantaneous communication of cell phones enables plans to be restructured up until the very last minute. This often means that friends will not know Friday afternoon what they are doing Friday night because they are waiting for a callback from others who are still deciding or revising plans and keeping options open. This practice of delay seeps over to more formal invitations to weddings and anniversaries and appears to be widespread.[6] A recent family wedding illustrated this pattern when a large number of invitees responded to a previously mailed wedding shower and wedding reception invitation only when they were called on the phone. We knew they were coming, but they had not mailed back the RSVP card. Explanations or responses when reached by phone: "Oh, yes, I just haven't gotten around to it yet. I had meant to. Oh, didn't I send it? Sure, we're coming, but we're just working out our schedules."

I've interviewed others who have organized wedding receptions and other social events where an accurate count of the number of guests is needed for the restaurant or hall. These informal surveys confirm my sense of a decline in the courtesy of a prompt reply. The planners had to make large-scale guesses as to numbers of guests because a substantial portion of (mostly younger) family members had not (yet) replied. The bridal family understood these friends or guests were indeed coming but had not (yet) gotten around to sending back the reply card.

Lapses in conventional manners are one of the small social changes that follow from new technologies and emerging social mores. The very real advantages of cell phone calls, texting, and e-mail invite adaptive, new behaviors. Some invitations now read, "Only if you are *not* coming, please reply."

There is no disputing the value of new instant communication devices, but it takes some time for society to smoothly integrate them and develop new rules of etiquette and of practical behavior. When my wife and I, on some shopping expedition, get separated in a large store or I'm browsing in a bookstore and she has completed her task, we can alert each other by cell phone. I've watched people in the supermarket check with a spouse at home to ask whether the spouse wants a large or small bottle of milk or bleach or whatever. No location seems off limits for the omnipresent cell phone. Cell phones break down barriers to communication by enabling one to stay in touch walking down the street, at the beach, at sporting events. But cell phone use in public bathrooms? That perplexes me, though I do understand health clubs' and gyms' rules against cell phone use in locker rooms because of concerns about photos being taken with smart phones. At concerts and the theater, public announcements asking us to turn off all cell phones and pagers have become so routine that not everyone pays attention and there is the occasional ring during the performance. The sight of people checking their phones or talking into their phone has become commonplace. A curious story I heard while researching this book was about a father and son who would go for a regular Sunday walk in the woods, both talking on their cell phones—talking to each other but separated by several miles. (This is a vivid example of how phones eliminate space and allow us to be in two places at once).

When new communication technologies appear, especially in the area of media, the law is slow to adjust. Thus, to boast on Twitter that one is going to blow up a bridge or the principal's office may be a joke or sarcasm in the eyes of a sophomore who sends the Twitter, but it could be a terrorist or dangerous threat in police eyes. In situations like this, even rules about how free speech is are not settled.

How Did We Get Here?

Since 1985 the number of cell phones has increased from a few thousand to more than 4 billion—nearly two phones for every three people on the planet, as estimated by the International Telecommunication Union in Geneva.[7] There are over two hundred million cell phones in the United States, again, nearly two phones for every three people in the country.[8] The different styles and expanded functions of these portable phones mark the sharp upward curve of the telephone's evolution, which serves as one gauge of the extent and speed of changes in communication and our behavior. The early standard AT&T land line phone, the squat, black, rotary-dial model, remained unchanged for decades until AT&T introduced the stylish Princess phone in 1959, a compact phone convenient for use in the bedroom because it had a light-up dial.[9] From the first telephone to the Princes phone was 85 years, there were 20 years from the Princess phone to the first heavy brick-shaped mobile phone, 10 years from the early cell phones to the smart phones like the BlackBerry, and 5 years from the BlackBerry to that marvel of multimedia functions, the iPhone. Now new generations of the iPhone seem to appear about every six months. These recent smart phones offer many additional features—GPS capability, cameras, media players—which are quite different from the original voice-to-voice function of the original phones. Today a customer needs an instruction manual to fully exploit the features and applications on these devices. Are they still phones? Of course, but they've changed and allowed us to do much more in our lives.

Addictive? Or Merely Lacking in Courtesy?

Almost all high school and college students use cell phones, and for some students, usage verges on an addiction. I have to repeatedly request students not to text during class. But increasingly such compulsive texting is not limited to teens and twenty-year-olds, nor is rude or thoughtless behavior limited to a younger generation.

Several colleagues tell me of instances where they are at lunch with a peer (in their forties) and the dining companion will pull out a phone and open to Facebook. When asked about this, the response is usually something like: "I thought of something and wanted to upload it while it was still fresh in my mind." Another response was, "My students do it, so it's very modern." Incivility is behavior that suggests one is unmindful of how one's behavior affects others. Electronic devices occasionally lead to actions that others regard as inconsiderate—rude—because of the technology's powerful ability to claim our attention, no matter where we are or what we're doing.[10] I have seen people accept calls while they were at a small dinner party. Attending to one's phone, whether to send or receive texts, while working or talking with colleagues or friends is certainly annoying, if not insulting, to those physically present. Some may rationalize texting and sending e-mails during meetings as a way of multitasking and being efficient—but it is really a way of splitting oneself off and vanishing psychically from the room. When people disappear from formal or informal meetings via their electronic devices, their colleagues interpret such behavior as a message: "You are less important to me than my cell-phone/P.D.A./laptop/latest gizmo."[11]

Does this suggest signs of addiction? People who are addicted to their cell phones find it hard to put it down or shut it off—especially in situations where its use is inappropriate (in a restaurant or a doctor's office) or dangerous (while driving). It is estimated that almost three-quarters of cell phone users talk while driving; this means at any given moment on U.S. highways, 10 percent of all drivers are on their phones.[12] We can pass over in perplexed silence (or to the sound of flushing water) the use of cellphones in a public restroom. People might be addicted to their phones if they express worry or anxiety about unable to make calls or text, or if they are constantly checking their phones for a missed message. High bills might also signal an addiction similar to gambling, especially if the cell phone user gives little thought to his or her incessant use.[13] In short, some cell phone users fall into the broad category of being addicts by their compulsive cell phone use. When an action is so reinforcing as cell phone response, our brains are pumping dopamine into our synapses for that gratification feeling that is not unlike taking a puff of a cigarette or a hit of cocaine. Something about receiving a text or a response to a text is reinforcing and rewarding. The dopamine makes it a feel-good, let-me-do-that-again behavior, over and over. In addition to dopamine are those natural endorphins that give us the "runner's high" feeling, and that kick when we feel pain.[14]

Conventions Surrounding Cell Phones

With caller identification commonplace on phones, new forms of greeting can be heard. "Hi there." "Hi Sally." Or the stripped down: "Hey." Or even "S'up." Simply answering "Yes" appears excessively curt and conveys the sense that the one answering is really not interested in a call at this time.

Social conventions—good manners, etiquette—get refined to lubricate our interactions in public. As technological innovations emerge, it takes a period of time for agreed-upon rules to become established—however arbitrary those rules might be. Thus when the automobile became commonplace, drivers worked out the pattern of keeping to the right side of the road. As communication technologies make their appearance, society is still refining the rules and, for the most part, we do not hear pagers and cell phones jingling in restaurants and theaters. Good manners frown on interrupting a conversation with the actual person one is with in order to take a call on a cell phone. States are passing laws prohibiting texting while driving, and sometimes even the use of a cell phone while driving.[15]

Changes Produce New Words

Instant messaging and texting introduce not only new behaviors but new terms and abbreviations. Texting itself has been made into a verb form, and IMing (instant messaging) is common. Most of the shorthand terms are amusing abbreviations of conventional terms. "Aml" means all my love, "B4n" bye for now, "Otoh" on the other hand, "Cu2moro" see you tomorrow, "Dk" don't know, "F2t" free to talk, "Jam" just a minute, "Ybs" you'll be sorry, "Lmk" let me know, "Lol" laughing out loud, "Omg" oh my God, and "Sit" stay in touch.[16] More curious are the novels published in Japan that have been composed on cell phones by authors who thumb in the text.

Our language is an important barometer of social and technological changes, because we need to name things. The field biologist Kate Jackson noted how jarring and rude it seemed when the people around her in the Congo bush country didn't say please and thank you when she expected it. Then she realized they didn't have the words *please* and *thank you* in Lingala. She offers the idea that in English we don't have a precise translation for *bon appétit,* a polite wish offered to someone when eating; we tend to use the French.[17] English readily develops new words to refer to new experiences and new aspects of technology, such as pixels, high definition (HD), random access memory (ROM), and gigabyte.

Words show shifts in thinking as well as new concepts and behaviors. Indeed, some words have had to be retroactively adapted to differentiate familiar objects from the new. Thus old watches with hands now are called *analog* to distinguish them from digital watches. We refer to *day baseball,* which is much rarer now than night games. *Regular coffee,* of course, refers to caffeinated coffee to distinguish it from decaf. *Paperback book* is a familiar term, but now we use the more explicit *hardback* to refer to what used to be called simply *books.* We distinguish a *business partner* from a *life partner,* and *fresh* pepper from *ground.* Standard transmission used to be the norm for cars, but now a car buyer must explicitly ask for it because automatic transmission has become far more common. We do need to know if a performance is *live,* and if turf is *natural.* Tap water lets us know it is not from a bottle, and land lines once were the only kind of telephone connections. A *personal trainer* differs from the medical trainer assigned to a team; a *personal coach* may refer to a therapist as opposed to an athletic coach.

New technical terms and acronyms emerge to give expression to new actions and capabilities, such as Wi-Fi and DOI (digital object identifier, used for citing and linking to electronic documents). And then there are the little *smilies* embedded in e-mails, such as ϑ or :) or ;-).[18]

Sexting

Cell phones that have the application to take and send photos enable sexting. Sexting is the sharing of nude or partially nude photos via cell phone. While it may be shocking, the practice of sexting is not unusual, especially for high schoolers around the country, although few of the teens who engage in this realize they are breaking the law. Television and newspapers routinely report on some incident when an inappropriate picture comes to the attention of a parent or appears on the Internet. In one such incident, police investigated reports that dozens of students at a middle school used their phones to circulate a nude photo of an adolescent girl.[19] Roughly 20 percent of teens admit to participating in sexting, according to a nationwide survey by the National Campaign to Prevent Teen and Unplanned Pregnancy.[20] Sending and posting nude or seminude photos or videos might start at a young age but becomes more frequent, it seems, as teens become young adults, with nearly a third of older teens or young adults reporting being involved in sending or posting nude or seminude images of themselves. Sexually suggestive messages (text, e-mail, IM) are even more prevalent than sexually suggestive images. About 50 percent of teens report they have sent or received such messages.

Voice-Recognition Answering Systems

When communication technology works well, it gives pleasure and makes routine business and daily life smoother. It is a modern frustration, however, when calling some organization to have one's call received by an automatic answering machine at a call center that uses voice-recognition software. When one's call is answered by an automatic phone system that works as designed, the arrangement helps speed the requests of callers. But glitches in the system seem to exasperate callers twice as much. We have all heard the warning that this "call might be monitored for quality assurance." Then there is the usual decision-tree menu of "Press 1 for English, 2 for Spanish" followed by "Press 1 for . . ." and so on. Often the menu choices do not quite match what the caller is seeking. The reader probably has endured frustrating calls during which he or she feels tense and angry while on hold and subjected to recorded music or advertisements. I have little success with computerized voice-recognition answering services that ask me to say my name or give identifying numbers. These computers usually do not understand my words, even though I am a native speaker of American English. "I'm sorry," the pleasant computer voice too frequently says to me, "could you repeat that?"

Fashion and Appearance

Ties, Sagging Pants, and No Hats

The interconnection of behaviors in our complex society was brought home to me when I was shopping recently. I was trying to find a shirt in my size, which should not have been difficult since I am quite average. But I noticed the sign listing available sizes: from medium to triple extra large. I asked the salesman why there weren't any smalls. He said, "Look around you." He indicated a number of other male shoppers who were more portly than myself. He told me that the store didn't carry many small sizes: "Try the boys' department." I've interviewed other store managers who concur that small sizes are less and less common. Indeed, glancing around, one does notice many overweight Americans. (In an article published in the *Journal of the American Medical Association,* the Center for Disease Control and Prevention says that about one-third of adults, as of

2008, were obese, but that the rate of increase for obesity in recent decades might be slowing.)[21]

How we clothe ourselves is a public behavior acutely attuned to the spirit and temper of the times. Premodern dress codes emphasized class differences, rather than gender roles. Males generally downplay the importance of fashion and seek to be restrained and sensible. Changes in men's fashion tend to be milder and less frequent than changes in women's fashions. Of course, men are influenced by fashion, but must appear not to be so, leaving it to women to be different and extravagant.[22]

When I started teaching in the 1960s, the dress code for college male faculty was jacket and tie. Then by the 1990s just shirt and tie sufficed, and sometimes the tie could be left at home. Recently my young colleagues began to appear in classes in jeans. My fashion eye has deficiencies, but even I recognized that jeans had finally climbed to a position of acceptability, if not respectability. Could high-end and prewashed jeans seem to be slumming a bit? For a brief period students were paying a lot of money for preripped and prestressed jeans. Faculty scientists who work in the lab tend to dress down. But I seem to be out of step and never in sync with fashion trends: just as I adopted casual slacks and open-collar shirt, I noticed some of my young male colleagues being daring and presumably cutting-edge by arriving for class, yes, now wearing ties. After decades of Casual Fridays, it may amuse or dismay aging men who hung up their ties to see that now sales of ties to young men aged 18 to 34 are on the rise.[23] When I interviewed a young biologist, he told me he always shows up for class in a tie—and blue jeans. In high schools and earlier grades educators need to be especially vigilant in balancing issues of freedom of expression versus what is objectionable and provocative. That is, some young females may expose excessive amounts of flesh and some young males may wear clothes that suggest gang affiliation. The style of sagging pants on males originated in prison where oversized uniforms were issued without belts so as to prevent suicide and their use as weapons.[24]

The casual look is clearly in—even in situations that used to be regarded as formal. Men in jeans attend symphony concerts. Jeans, no longer associated with the working class, have migrated to the status of trendy and established. Upscale and carefully distressed jeans now can be read as a fashion statement, perhaps as a costume of "people eager to communicate their indifference to appearance."[25]

During a cold New England winter day, a day on which I require a hat to keep warm, I observe that almost half the passengers on my T ride are hatless. Well, what about summer? On blazing hot days, I notice a similar percentage who are bareheaded, apparently not needing protection from the sun. Pictures of men attending baseball games or heading off to work during the 1930s and 1940s show almost all of them wearing lightweight straw boaters or dark felt hats. Styles change. It was once thought that President Kennedy, by not wearing a hat to his inauguration, destroyed the habit of men wearing hats. More likely, he was simply one of millions of men who slowly got out of the habit—except for the newer practice of some men wearing a baseball cap inside a diner or a casual eatery.[26] To see a man wearing his hat in a formal, white-tablecloth restaurant is more puzzling. Possibly wearing a cap has become a habit, akin to putting on a piece of jewelry or some other fashion accessory, and thus the traditional conventions don't apply in this informal culture. Women, who tend to be more attentive about dressing behaviors, currently have abandoned the fashion of wearing the stylish (sometimes, outlandish) hats of several decades ago.

Male fashion tends to be conservative and to change far less than women's styles. The business suit remains the formal signal of one's seriousness and reliability. Banks and financial institutions still expect their male employees to be conservatively dressed so as to reflect the solidness of those institutions. Are the gorgeously dressed male television sports commentators setting a new trend?

Only with timidity do I make the most cautious judgments of female styles except to notice a considerable range within the ranks of my women colleagues. Senior women faculty and administrators tend to mirror my sense of what is professional, while younger female colleagues often resemble the undergraduates. Freshmen female students are the most accurate barometer of the newest popular styles, and I have now developed a sense of the year-by-year rhythm of change in youthful styles, perhaps more cleavage this year or perhaps more midriff exposed.

Such changes in fashion and cultural rules are common. Such changes are most noticed when placed side-by-side, for comparison, with alternate forms. For instance, I am still startled at seeing some of my scruffiness-prone young males decked out in tie—and even a suit, perhaps—as they brush and polish themselves for internships or job interviews. In a recession, more conservative dress conventions return, especially as jobs become scarcer and job seekers have fewer options. There is also some circling back of conventions, and older men will remember the bedraggled jackets and askew narrow tie as the uniform of their undergraduate days.

As the culture changes rapidly I find myself sometimes a bit anxious when formerly clear rules are neither clear or consistent. I commonly check with my wife about events for which I am doubtful: should I wear a tie, a tie and jacket, or can I go business casual?

Changes in social structures and rules tend to favor youth—especially when there is no venerable tradition to provide guidance. Unencumbered with ideas of the past, youth seem to glide easily into the modern world of technology and rapid social change. Middle-aged adults who fail to appreciate the values distinct to each developmental period often resort to clinging to vestiges of their own youth—aided by technologies of Botox and plastic surgery. Sales of hair coloring remain steady amidst other social changes.

What about unexpected consequences in fashion? Apparently young children commonly wear shoes and sneakers that have Velcro fasteners. This is convenient for child and parents, but seems to delay learning the skill to tie simple knots.

Tattoos and Piercings

Tattoos now peek out or sometimes blaze forth from various body parts of men and increasing numbers of young women. Body piercing as a fashion statement seems to have crested, but tattooing as a behavior that used to be at the edge of conventional society has moved to the center and become an accepted fashion statement. *Life* magazine estimated in 1936 that 10 million Americans (about 6% of the population) had at least one tattoo.[27] The website on which that information can be found, VanishingTattoo.com, gathers statistics from the Harris Poll and the American Academy of Dermatology to estimate that now more than 16 percent of Americans have at least one tattoo—that's more than 40 million people. Most of the increase in numbers occurs within the 18–40 age group, in which more than one out of three have at least one tattoo. The Pew Research Center reports that 30 percent of this same cohort also has a body piercing somewhere other than the ear.[28] That, of course, refers to piercings in the nose, lip, belly button, or more private places. Clearly many young people are turning to some form of body art as an important expression of their modernity and individuality, as well as yet another way to quietly rebel against older standards. Popular television shows like "Miami Ink" and pictures of tattooed professional athletes provide cultural modeling and permission, if any is needed. In my health club's locker room, when I showed interest in the elaborate and colorful tattoos on the arms and back of a buff former Marine, he so enthusiastically explained his "pieces" that I could almost begin to think about what image or expressive phrase I might choose to express the essence of my identity.

Greater Emphasis on Fitness

People have competed in sports and games for millennia, but the British leisure classes established rules and gave impetus to some of the games that entertain us now, such as soccer, cricket, baseball, and tennis. The American concern for health and fitness fostered the relatively recent surge in new health clubs and greater participation in sports. Title IX of the Education Amendments of 1972 lowered barriers for women to participate more fully in competitive college sports.

One manifestation of the fitness trend is the appearance of women lifting weights and working out. Of course, women have been caring for their health for a long time, but I notice the changes by their presence in health clubs and by their participation in competitive—even contact—sports. Title IX has added more women's teams to college sports and encouraged girls who wish to play what used to be regarded as boys-only sports, such as wrestling, ice hockey, and lacrosse. Wrestling may be the ultimate contact sport, and "it can be a startling sight, teenage boys grabbing girls' thighs, girls straddling boys, boys riding girls' backs and trying to flip them onto their backs."[29] For the most part, girls who want to wrestle must practice with, and compete against, boys. Nationwide, about 5,000 high school girls wrestle, according to the National

Federation of State High School Associations. Women's wrestling is now an Olympic sport.

Sports fans have adopted new behaviors. Increasingly television emphasizes sports, and more youth participate in sports. Fans increasingly dress in the branded clothing of their teams or display the name of their athletic heroes on their shirts. Stores spring up to supply specialized athletic equipment and clothing with team logos.

Technology has greatly altered the sports of tennis and golf. The new racquet technology of the last two decades provides a large sweet spot in each racquet's head that allows average players to hit the ball with increased pace and accuracy. Younger players learn to adjust their grip, which allows for more topspin on the ball. This forces a player to start thinking differently about tactics. The speed of the ball in professional tennis is so quick that fans sees very little serve-and-volley style of play; players stay at the base line but rip powerful ground strokes. In short, the new racquets changed the tactics and overall style of tennis, requiring the professionals to be faster and stronger than ever before. Women's tennis attire reflects the greater athleticism need for the modern game, and women compete at far more intense levels than their mothers did just two decades ago. Formerly an activity dominated by decorum and proper dress, tennis has taken on more democratic values. As a tennis player and fan of televised tennis I've noticed in recent years the new behavior of fans at the U.S. Open who shout encouragement from the stands. Even at traditional Wimbledon, which still requires white clothing for its players, fans have been observed to do the Wave—the synchronized standing of cheering fans that is usually an expression of enthusiasm more common at U.S. football games.

Aluminum bats changed baseball, at least at the little league level. There is a good reason why professional teams use only wooden bats: the ball comes screaming off the metal bats when used by better players. Baseball fans know that changing the rules (adding the designated hitter in the American League) changes the style of play. Professional baseball players used to have knee-high stockings, but now commonly let the legs of their uniform pants droop around their ankles. Basketball players have allowed their uniforms to evolve also, with their uniform shorts being long enough to cover their knees.

Other Public Behaviors

On my commute to work, I pass by a commercial building, outside of which I see small groups huddling over their cigarettes. When I look for changes, I find it often in the small details of life that easily can get lost amidst the clutter of daily business. Thus, I cannot but help notice small groups of smokers clustered around entrances of buildings as they puff on their cigarettes. I feel badly for them in winter time because they often stand in the cold without jackets. Not that long ago one had to request a smoke-free section of the restaurant ("Smoking or non-smoking?"). Missing also, but scarcely noted, are the cigarette dispensing machines formerly located in many restaurants and most bars.[30] Some changes show progress.

Some behaviors are undoubtedly due to people untrained in traditional decorum or practices. Theater devotees say manners are breaking down faster than ever among audiences. On Broadway the veteran singer Patti Lupone broke character in *Gypsy* to scream at an audience member taking pictures—which the program notes clearly is illegal. Lupone said that for months she had been pointing out to ushers audience members texting or taking pictures with cell phone cameras. During the nude scene in *Hair,* members of the audience were snapping pictures despite warnings not to do so. Audience members arriving late asked the actor to wait. During a recent quiet moment a woman in the mezzanine screamed, "How 'bout those Yankees!" One theory as to why audiences sometimes act as if they are in their own living rooms suggests that Broadway is vulnerable to boors because it is under pressure to fill seats in a recession, and that theater audience demographics are changing to include more children, teens, and other people who don't usually go to the theater.[31]

The Media and Changing Social Arbiters

What are some of the social engines that drive new behaviors? When Clark Gable took off his undershirt in the 1934 film *It Happened One Night,* supposedly T-shirt sales sank. (The online myth buster, Snopes.com, is dubious about such a cause-effect relationship in this instance. The website finds this incident to be yet another instance of a common phenomenon: popular figures who reflect shifts in societal norms that are already under way. So possibly Gable with no undershirt and President Kennedy with no hat were following trends, not starting them.)[32] Generally speaking, however, people do look to trend-setters for fashion tips and emerging styles. Consider the fashion statement embraced by many men of sporting a two or three days' growth of beard: are the male models and celebrities starting or following this trend? Do professional athletes with prominent tattoos start or follow a trend, or make a trend acceptable? Did celebrity females like Angelina Jolie and Lady Gaga give impetus to the trend of females sporting tattoos? Online websites give some insight to how two celebrity females think about this contemporary form of body ornamentation. In Jolie's case, a tattoo on her left shoulder blade is the Japanese symbol for death and it reminds her to live. She had it removed and now has the Khmer tattoo for her son Maddox in its place. A tattoo on her right arm is the Japanese symbol for courage. A cross tattoo that covers the little dragon with the blue tongue was done in Amsterdam. Next to it is a phrase in Latin: "Quod me nutrit me destruit," meaning "What nourishes me, destroys me."[33] Similarly, Lady Gaga tells an interviewer that the tattoo on her inner arm is a tribute to her "favorite writer, Rainer Maria Rilke. . . . In German he writes, 'Confess to yourself in the deepest hour of the night whether you would have to die if you were forbidden to write. Dig deep into your heart, where the answer spreads its roots in your being, and ask yourself solemnly, must I write?'"[34] Explaining that the tattoo on her shoulder was done in Japan, she explains that there was a collaboration with the legendary Japanese photographer Araki. "I was bound by Araki's personal bondage artist, by several ropes and Japanese knots. Araki photographed me, using a series of several cameras. He did not photograph my image; he photographed my soul."[35]

How New Behaviors Spread

One way to try to explain the rapid diffusion of a new behavior is to see it as a form of communication. Thus, many people quickly began to adopt cell phones because they saw others using them, they saw advertisements, and they saw information about them in the media. Everett M. Rogers describes stages in the diffusion of innovative products and behaviors.[36] "Innovators," probably 2 percent of the population, are daring and willing to take risks. They are the ones who conceive and develop new ways of doing things and new behaviors. The baton is picked up by *early adopters,* about 14 percent of the population. Then the crowd follows, with the *early majority* making up about 34 percent of the population. They are followed by the *late majority,* who join the party after it is well along. The *laggards,* or Luddites, about 16 percent of the population, are slow or reluctant to embrace new products or behaviors because of disinterest or financial constraints. They may even fear new technologies or behaviors, especially if they feel threatened or are worried about losing their jobs.[37] When we plot the frequency of individuals adopting a new behavior or new technology, the resulting distribution is an S-shaped curve. That is, in the beginning, only a few individuals adopt during a particular time period; then the diffusion curve climbs as more and more individuals adopt the behavior or innovative technology. Eventually the curve levels off and flattens as fewer and fewer individuals remain who have not yet adapted, and the diffusion process is finished. This process can be seen again and again, whether it is with the introduction of new consumer items like televisions or cell phones, or with new behaviors like cyclists using helmets or farmers using highbred corn seeds.[38]

Obviously some new trends will enter into the mainstream and others will flicker out. Marketers seek to foster widespread adoption of products, and research the process of what makes a hit or creates buzz, that is, people excitedly talking about a product or fashion or show or book. Marketers will try to manipulate the process, perhaps by actively seeking out early adaptors or *influentials* and providing them with early versions of the new products. Influentials or early adopters are the ones who first buy, or buy into, an idea or fashion or product, and tend to be the ones to spread by word of mouth and create buzz. Every behavior and product category contains these initiators who spread the word or the behavior. Finding and seducing this important group is an essential step in creating a trend that then spreads to the larger population, the masses.[39] Successful products and fashions and behaviors have what is termed *legs* and will last. They fulfill some need in people or provide status or pleasure in some way.

The consequence of innovation and new behaviors is the change that occurs. Change can be desirable or undesirable, functional or dysfunctional, or combinations of both. New behaviors and the innovation that "occurs so frequently in modern society"[40] create a climate of uncertainty as well as a host of benefits. The role of manners and social rules is to smooth daily life and make it somewhat predictable. Rituals can help us to get through awkward or difficult situations. Increasingly, however, people feel unsure of proper protocol and don't have confidence as to what the social rules are. A strength of the American culture is its tolerance of the new and the informal. This has contributed to a sense of freedom and openness as well as uneasiness and frequent lack of social clarity.

Contemporary guides to proper behavior and etiquette concur that "the modern explosion of social complexity" leads to uncertainty about the rules.[41] It is not so much an issue of whether people have gotten ruder; as the times have changed, the country is more fragmented, with increased diversity, different priorities, values, and experiences.[42] Things are different. As I suggested in chapter 4, people feel a strong sense of time pressure. As I will suggest in the chapter on family life, parenting in particular has become more complex, with parents much more engaged and involved in their children's lives, soccer practice, monitoring computer use, and so forth. Every culture seeks to make sense of the world around it and evolves practices or rituals for the transitions of life, such as birth, marriage, graduation, and death. "But people don't all react to change in a coordinated fashion. . . . Everyone processes changed at different rates. This means that for any situation there are probably quite a few versions of 'correct' behavior to choose from."[43]

One cause for some decline in politeness is that life is so hectic and people are so busy and pressed for time that they forget to be polite. In a survey about rudeness, people sought for explanations why there may indeed be a decline in civility. One plausible explanation was a sense of overload. "People simply have so many obligations and so much on their to-do lists that basic courtesy is often an afterthought. Nearly half (47%) believe a major reason for disrespectful behavior in our society is that 'life is so hectic and people are so busy and pressed for time that they forget to be polite.'"[44] More than half of those surveyed admitted that sometimes they were so busy and pressed for time that they were not as polite as they should have been. One Florida man said, "People want to pack as much as possible into 24 hours. The day hasn't grown, but people want to do more and more and something has to give."[45]

Many new behaviors spring up in times of change. In exchange for the loss of some "of the ways things once were," we get new patterns. I do *not* miss the blue laws that kept Sunday a very closed down day of the week; I like the convenience of opened stores. I like the greater informality of many official events that allows for more casual dress; although sometimes it is nice to really dress up and have a secure sense of knowing what to expect.

6

Faster and Faster Time

In Time

I savor the temporal unreality of the last two or three minutes of a professional basketball game. A referee's whistle or a coach's "Time out!" can magically stretch 2 or 3 game minutes into 10 or 15 actual minutes. I wish I could be like a coach who could call a time out in life, and stop time when I want more time to think about or savor a situation. Much of modern life has the rapid pace and pressure of a basketball game, without the magic of the clock slowing or stopping.

Is time speeding up? A common response of friends and colleagues to my queries about how things are going is, "Fine, but I'm so *busy*—I don't have enough time." Time has speeded up for many of us and become more scarce. Time metaphors and expressions abound. Time we saved gets *spent* quickly. The language of time suggests it is very much on our minds: *on* time, *free* time, *real* time, the greatest of *all* time, *over-* time, *part-* time, *out of* time, *killing* time.

This chapter considers several reasons for these changes in our experience of time. The chapter also examines how we use time and how the technological forces of modern times speed up time and change us.

Constant Reminders of Time's Passage

Messages from many sources remind us of the rapid passage of time: fast food, instant breakfast, the fast lane for the turnpike, speed-dialing for our cell phones. Talk radio announcers commonly say, "We just have a few seconds left," or "Make your comment brief, as we are running out of time." Psychotherapists, trained to be disciplined about the length of therapy sessions with clients, use the phrase, "We are out of time." Each month the notice of a mortgage payment due or a credit card bill reminds us of deadlines and of a moment in time by which I must accomplish something—even if it is just to drop a payment in the mailbox or access the simplicity of online payments, which speed up the process even more. Digital cameras have rendered one-hour photo processing obsolete.

The Internet and communications technologies especially bring us in touch with speed. Each Google search rewards us with thousands or millions of "hits" in a fraction of a second. (I used Google to search for "How to Slow Time," and was rewarded with 123,000,000 hits in .34 seconds.) E-mail and cell phones shrink space and compress time. E-mail was initially thought to give us a breather from needing to reply instantly to telephone calls, while being faster than the U.S. Postal Service—snail mail. But e-mail speeds up the process of communication, enabling us by rapid back-and-forth communication to resolve issues far faster than paper letters back and forth. A director of an education program told me it used to require weeks for him to resolve certain issues with foreign countries. For example, he said he used to need three or four weeks to send and receive correspondence with New Zealand; now he completes his business within an hour electronically. Cell phones provide even greater, quicker access to a person, no matter where they are. Phones shrink distance and further accelerate communication.

Computers especially, at which we sit for several hours each day, remind us of speed—or its lack—and time's flight. My computer's processor is very fast, so much faster than my previous computers. Yet how quickly I get impatient when something is taking a few extra seconds to download or boot up, when the Internet is slow, or when my security software is slowing down my start-up time. Early computers tended not to make us impatient, because we were new to the technology and dazzled by whatever speed the computer used.

Basketball, football, hockey, and soccer are played by the clock, so that even in our entertainment we are reminded of the tyranny of time. Older sports, like baseball, cricket, and tennis are played by the inning and the match rather than being timed. New improvements of electronic timers and digital clocks visually show us the seconds flickering away—especially in track or swimming or downhill skiing—to the hundredths or even thousandths of a second.

The Experience of Speed and the Perception of Acceleration

When we speed up the present, it changes how we think about the past: it slows down the past. Historian Steven Kern gives the example of a man riding a horse for 20 years; but after the auto is invented and the man travels by auto, he experiences both acceleration and slowing. The very acceleration of the auto transforms his former means of traveling into something that had never been slow, but up to then had been the fastest way to go.[1] Most of us experience this sense of change when we get a new computer or a new cell phone. Improvements in apps and processing speed make our old models appear to have been slow, whereas when they were new they were fast.

Impatient myself with my computer's start-up or linking with the Internet, I notice more frequently now others' restlessness or boredom in situations when the pace of an expected activity is seen as slow. Students grow up in a world of television and fast computers. Consequently teachers try to make classes brisk and entertaining by using PowerPoint and video clips. Professional conferences similarly have an up-tempo style to keep audiences engaged and focused. We have developed an appetite for interactivity and being stimulated.

We have become accustomed to the speed of transportation, with jet travel and with automobile traffic that frequently goes 80 miles an hour—well above the speed limit. Time-saving devices like dishwashers, washing machines, and automatic coffee makers surround us. But we have so well integrated them into our lives that long ago we began using the time they saved us for other tasks and activities.

And we have gotten so used to the speed of communication by now that we scarcely appreciate that a phone enables us to communicate over vast distances at seemingly instantaneous speed. If we are moving faster (via car or jetliner) and have more sensory impressions via television and computers, than our general sense of speed is increased because of the increase in stimulation and rapid flow of images. We are surrounded by machines that are very quick—the computer's processor, for example. We can take digital pictures and have the pleasure of seeing them immediately or of printing them ourselves—no more need of the One Hour Photo store.

The perception of time's speediness is further heightened by change—that is, the appearance of what is different accelerates the sense of time passing: the more change there is, the quicker time seems to be. In other words, the more busy I am, the more quickly time seems to slip past. The more I divide time (my schedule, my week, my day, my hour) into smaller chunks, the more rapidly the minutes speed along. I tend to chop up my tasks in order to stay focused and keep my concentration, but the more I divide my time into small chunks, the more rapidly time seems to speed along—and the more fragmented my attention is.

Time is money in our culture—and we don't waste money or time. One of the inventions of modern life is the billable hour. Lawyers and psychiatrists have carved up the 60-minute hour into lengths of time on which they put a price or for which they charge an insurance company.

Age contributes to the sense of acceleration. In my chronological age, each additional year is a smaller part of my whole life and thus a year seems shorter than when I was two, and a year was half a lifetime; at 60 a year is one-sixtieth of one's life.

More Information Makes Time Go Faster

When information was rare or hard to obtain, it was of value. We now have many channels for receiving information: e-mail, the Internet, cell phone, news updates on CNN and radio, multiple television channels. As we produce more information and increase our many channels of communication, the relative value of information declines. More information crowds into the same box of time. More information makes time feel crowded and accelerating. As more and more information from any source is consumed, more activity is pushed into available time. Thus, if I resolve issues rapidly by e-mail and cell phone, I am saving time on decisions and projects that used to drag out because of delays in getting responses from others. I now have more time, which I tend to fill by taking on more projects or looking up more information.

Author T. H. Eriksen comments, "More and more activity is squeezed into every free moment, and the gaps in time and our schedules are filled."[2] New technologies help make us more efficient. When we are waiting for a bus to arrive, when a meeting gets cancelled, when a flight is delayed, we tend to fill the gaps in our time by a quick phone call or sending a text or reading. The net result is a culture that is quick and momentary, rather than slow and deliberate.[3]

A growing number of messages fight over a forever shrinking vacant space in our daily lives. One consequence is that each of us gets to spend a decreasing amount of time on each piece of information.[4] When there were fewer available television channels, people probably watched programs in a calmer and more continuous mode. Faced with many more available television channels, many of us surf restlessly with the remote control.

Many of us cram more into the metaphorical box of an hour, seeking to do more, to multitask. As time becomes compressed, it seems to pass more quickly. People recently thought that e-mail gave us a breather from needing to reply instantly, the way we do to a telephone call. Of course, e-mail is much quicker than snail mail. But with smart phones and BlackBerrys, people now get alerts to an e-mail and they can respond almost as quickly as if they spoke on the phone. Is time being saved? Are people more efficient? Probably, but the trade-off appears to be the acceleration of time and the attempt to do more in time.

Television Slices Up Time

Television contributes to our sense of time accelerating. Watching television is to watch time being sliced up into smaller fragments. I especially notice this in commercials, which commonly last 20 seconds, 3 per minute. The flicker of almost one image per second creates a kind of visual excitement that gets my energy up and I'm guessing catches most viewers' attention. The frequency of images and quick cuts provides stimulation and a sense of speed. I've watched old films on the Turner movie network and even watched old ads on the Nickelodeon. Black-and-white films from the 1930s are paced much slower, having perhaps just one or two image changes per minute. Old television ads from the 1950s are agonizingly slow to watch. An Alka-Seltzer ad from that era that might last a minute or two seems to drag on much longer because of the dull—by today's standards—cutting and pacing. Advertisers on television buy segments of 15 seconds or 20 sections to show a very compact ad. But running more ads contributes to the sense of speed because more information is filling the same amount of time.

Television plays a unique role in structuring daily life. Television is "the one technology that is so clearly linked to time. It is accessed almost instantly, provides regular programming geared to the hours on the clock, and provides anchor points in time around which other activities can be regulated."[5] Criticism of television has been a popular sport for decades, but the reality remains that from its very beginning television has been a sales medium showcasing new kinds of life and leisure experiences. It promotes the notion of a quick fix, with problems being overcome in half an hour. Things happen quickly on television because advertisers want the broadest possible audience with short attention spans. It works against the traditional middle-class values of deferred gratification, and suggests that Americans can have what they want, now.[6]

Television and the Internet hasten time. Television airs interviews with foreign correspondents halfway around the world or even from a battlefield, so that time differences are blurred and everything seems simultaneous. Embedded combat reporters and camera personnel can take videos on a battlefield and transmit the report from a laptop using a phone-book-sized satellite modem called an RBGAN (Regional Broadband Global Area Network) to NBC or CBS's computer server and the report can be on the air very shortly after the action.[7] The blurring of time differences in a live-TV feed from abroad makes us feel that all time is the same, simultaneous. I watch a baseball game at night and turn off the television at 11:00 P.M. and when I pick up the paper on my steps in the morning I can read about the ending of the game. During the 2008 Summer Olympics, between the moment when Natasia Liukin won her gold medal and minutes later when I checked the Internet, her Wikipedia entry had already been updated. That's fast. Sports fans can watch a game on television and almost immediately go online, via phone or computer, to find commentary, other scores, even video highlights. Even though much of the information on the Internet is scarcely necessary for our lives, it is amazingly and quickly available at our fingertips.

What Is Time?

Time is the counting of motion. That is, time bestows a number on change. As I look at a clock, it marks the passing of time by pointing to a new number. An analog clock marks time by moving the clock hands past the numbers on its face; the digital clock counts the passage of time by steadily changing the number on its face. If there were no change to be marked, no forward motion of reality in time, there would be no time or time would stand still.

There are different kinds of time, such as objective time, subjective time, slow time, leisure time. But first we need to see how our modern sense of time came from innovations in technologies.

Technology Changed Time

In premodern times, natural rhythms and daily tasks organized the day, rather than clock time. Thus, it's dawn so it's time to get up, the cows are lowing so it's time to milk them, I'm hungry so it's time to eat the big meal of the day. Mechanical clocks began to spread in the 14th century, and by the 16th century most towns and parishes in Europe had a church clock with at least an hour hand. Minute hands did not become widespread until the late 17th century,[8] and their presence made timekeeping more precise and speeded up the sense of time. Clocks and watches externalize time in a similar way that writing externalizes and makes tangible thought. Time becomes something existing independently of human experience.[9] Pocket watches did not become widespread until the 18th century, and were common by 1900. The profusion of watches was both a response to, and a cause of, attention to short intervals of time and a heightened sense of punctuality.[10]

Until 1883 the entire United States operated on a chaotic system of varying local times determined by the position of the sun. In New York, local time varied as much as a minute from one side of the city to the other.

"Around 1870, if a traveler from Washington to San Francisco set his watch in every town he passed through, he would set it over two hundred times. The railroads attempted to deal with this problem using a separate time for each region," says author Steven Kern.[11] The requirements of the railroads (a growing continent-wide technology) led to the regularization of time, and geographic time zones were adopted by railroads on October 11, 1883. Once Standard Time was introduced, railroads continued to synchronize timekeeping and make more extensive use of accurate timepieces. Railroad companies, therefore, brought about the standardization of time. Governments finally got involved, and the 1884 International Meridian Conference in Washington, D.C, established Greenwich, England, as the zero meridian, determined the exact length of the day, and divided the world into 24 time zones of one hour each.

Increasingly in developed countries, and especially in cities, time was emancipated from natural cycles. Time structured the day. By structuring the day, I mean that clock time determines when activities occur rather than the activity itself; thus work starts at 9:00 A.M., there is a meeting at 10:00 A.M., class starts at 10:30, the ball game begins at 1:30, and my favorite TV show begins at 7:00 P.M.

Thus, modern time began: time made fractured and discontinuous versus the characteristic premodern experience of time as flow and duration. Another way of naming the change is *the primacy of clock time* versus *time in the mind as experienced by the individual.* The day was now fragmented primarily by clock time rather than by activities and natural cycles.[12]

The Phone and Time

The telephone destroys distance and introduces simultaneity. The telephone lets a person be in two places at once, and abolishes temporal and spatial distance.[13] The phone pierces through walls of privacy and—if we allow it—is ubiquitous; it can find us virtually anywhere, on the beach or in the anonymous city. This access gives immediacy to a person, breaking through barriers of public/private self. It was during the Second Industrial Revolution, during the decades of the 1870s through 1910, that a series of technological innovations (telephone, gramophone, box camera, the airplane, and the automobile) obliterated space and transcended time.[14] These technologies captured the instant (be it an image or a sound) or moved actual people or just their voices across immense distances at great speeds. Edward Muybridge froze the movement of people and horses by means of photography. Alexander Graham Bell's recording of sounds froze human speech.

Using these technologies changes our sense of time. Simultaneity suggests that the present is no longer limited to one event in one place or limited to local surroundings. By means of electronic communication, *now* becomes an extended interval of time that can include events around the world. As an experience that has spatial as well as temporal aspects, simultaneity has an extensive impact, since it involves many people in widely separate places, linked in an instant by the communications technology. One thinks of broadcasts worldwide of the World Cup games, or New Year's Eve celebrations, which have become so routine that we don't think twice about them. Breaking down barriers of distance and time, communication technologies make all places equidistant from seats of power across social strata, and hence of equal value. Protocols of invitations and appointments become subordinated to instantaneity. Telephones penetrate all barriers, except perhaps churches.[15]

Measuring the Increased Pace of Life

The pace of life is increasing and people are moving faster. The American psychologist Robert Levine devised three indicators to compare the tempo of life in different cities. Researchers measured (1) how fast pedestrians walk 60 feet, (2) how speedily a postal clerk fulfills a standard request for stamps, and (3) the accuracy of 15 randomly selected bank clocks (presumably to determine how important clock time was to a city's residents). From these observations the researchers suggest five factors that determine the tempo of cultures: "People are prone to move faster in places with vital economies, a high degree of industrialization, larger populations, cooler climates, and a cultural orientation toward individualism."[16]

Comparing the results with the original work done in 1994, a follow-up study showed that people were, on average, now walking 10 percent faster. Researchers in August 2006 found a busy street with a wide pavement that was flat, free from obstacles, and sufficiently uncrowded to allow people to walk along at their maximum speed. They timed how long it took

35 men and 35 women to walk along a 60-foot stretch of the pavement. They only monitored adults who were on their own, and ignored anyone holding a mobile telephone conversation or struggling with shopping bags. Singapore came in as the fastest city, followed by Copenhagen, Madrid, and Guangzhou in China. The only American city was New York, in eighth place. The key conclusion is that "the world is speeding up," says Professor Richard Wiseman, the British psychologist who headed the study. "This simple measurement provides a significant insight into the physical and social health of a city. The pace of life in our major cities is now much quicker than before. This increase in speed will affect more people than ever because, for the first time in history, the majority of the world's population are now living in urban centres."[17] People who walk fast are also more likely to speak and eat quickly, wear a watch, and get impatient, the study suggests. They don't like to sit still, sit in traffic, or wait in queues. The researchers believe that technologies drive the increased pace of life, especially the technologies that keep people constantly in touch with each other.[18]

Modern timekeepers can track time in small and smaller increments, which departs further and further from our everyday experience. The margin of victory in an Olympic swimming or luge event might be a hundredth or even a thousand of a second. To obtain measurements such as this, technicians have to devise more sophisticated tools. Thus, within a millisecond (one thousandth of a second) a bat presses against a ball, a bullet enters a skull and exits, or a rock plunges into a pond and its splash pattern pops into existence. Now lasers and special photography enable researchers to freeze time at a fraction of a nanosecond (one billionth of a second), which renders baseballs, bullets, and water droplets motionless.[19]

Such precision matters for worldwide communications systems. It matters for navigation by Global Positioning System satellite signals. An error of a billionth of a second means an error of about a foot, the distance light travels in that fraction of a second. Cellular phone networks and broadcasting transmitters need ever finer timing to squeeze more and more channels of communication into precisely calibrated units.[20]

Measuring Computer Speed and Capacity

Information and data get linked with speed and ever greater acceleration. Thus when we speak of the technology of information, communication, and computers, the amount of information goes hand in hand with speed, either speed of the computer's processor or of the rate of transmission.

The speeds and memories of modern computers and communication devices get expressed in exotic language and involve numbers that defy everyday understanding. A "bit" is either 1 or 0, true or false, and 8 bits make a byte (from *b*inar*y te*rm), a unit of computer information that is used in describing storage or memory capacity. Computers now have memories that are measured in kilobytes (1,024 bytes) which is equivalent to a page of text, or a megabyte, which is equivalent to a small novel or the contents of a 3.5-inch floppy disk. A high-resolution photograph contains about 2 megabytes of information. Five megabytes of capacity would be necessary to store the complete works of Shakespeare. One gigabyte (which is a billion bytes) is sufficient storage capacity for a movie at TV quality or a high-fidelity recording of a symphony. Two gigabytes of memory would be necessary to store about 60 feet of shelved books, while 2 terabytes (which is about 2 trillion bytes) is sufficient to store an academic research library.[21]

Ten terabytes of memory capacity could store the printed collection of the U.S. Library of Congress (29 million books). The entire collection (all books, 2.7 million recordings, 12 million photographs, 4.5 million maps, and 58 million manuscripts) would require 10 petabytes, which is 10,000 terabytes or 10,000,000 gigabytes.[22] How fast could this be transmitted over a digital network?

The speed of a microprocessor and the speed of data transmission is measured in gigahertz. One gigahertz (GHz) represents 1 billion cycles per second of processing or the transmission of 1 billion bits per second.[23] Thus, the contents of the entire Library of Congress could theoretically—if it had all been scanned into the memories of fast computers—be transmitted in under an hour. This rate of data transfer is equivalent to downloading three full DVD movies per second.[24] This estimate was made on the basis of a California Institute of Technology supercomputer transmitting at a rate of 101 gigabits per second.

For most of us who are not engineers, these speeds and numbers contribute to a sense of alienation from the marvelous devices we use daily. When these devices work well, they are marvels. Often, when something goes wrong with our software or some type of system failure, these technologies frustrate us greatly. When they work well, they vastly contribute to our information sharing and collaboration, on a scale and with a quality previously unimaginable. Holding an Apple iPhone 4[25] is to hold a slim phone half the thickness of a deck of cards that has the storage capacity of 16 gigabytes and speeds of 2.4 gigahertz—far more powerful and far quicker than the computers that were used in early space launches to the moon.

Different Kinds of Time

Real time is a current phrase. It is a *retronym,* an old reality's revised name that has become necessary to distinguish it from an innovation. Thus snail mail is distinguished from e-mail, acoustic guitar distinguished from an electric guitar, rotary-dial telephone from a touchtone phone, decaf from caffeinated coffee, and so forth. As technology enables us to manipulate time by recording images and sound and by communicating rapidly, real time is different from virtual time or prerecorded time. Thus real time, as author James Gleick terms it, is a "now-ness intensifier."[26] Your stock broker executes your order in real time, that is, quickly or instantly. A televised football game is in real time, if it is live, even though it has many instant replays (repetitions of time events), which blur real time and virtual time.

One of private, or subjective, time's qualities concerns our level of engagement. When we are thoroughly immersed in a pleasant activity, time seems to pass quickly. When we are bored and not actively involved, time seems to drag. When stimulation holds our attention, there is an impression of increased speed, of time passing rapidly. Accelerating time feels as if it is cannibalizing the slow time of family life. Are friends or family members on their cell phones or at their computers or on Facebook rather than attending directly to each other? People speak of spending *quality time* with children and partners in place of more abundant time with them.[27]

But what about leisure time?

The Work-Time Paradox

Americans feel more rushed now than a generation ago, even though time studies suggest Americans have gained about an hour more free time per day through time-saving devices. Where has all the free time gone?

Virtually all studies show that the average amount of time Americans work and have for vacation has stayed substantially the same for the past few decades. Yet more Americans than ever before feel squeezed for time.[28] How can this be? Two or three root causes present themselves. We are doing more in the time we have—time itself has not changed, it just *seems* to be accelerating. We have the technologies to do more *at the same time*. The technologies I have talked about in the last two chapters contribute to squeezing more into time gaps, to multitasking more. What are the activities that people are doing in addition to what they did previously? We have many more sources of entertainment. We stay touch with people though more channels, such as e-mail, Facebook, and cell phones. Parents are more involved with children—and kids' lives, as I shall show in the next chapter, are more structured and demand more of parents.

The structures of families have changed. Forty years ago only about a third of families had two adults working; now that figure has doubled to almost two-thirds. More women are working, while continuing to be actively involved at home. Work has expanded into nonstandard work-shifts, and thus work increasingly takes place at times that were considered private time or not work times, such as nights or weekends.[29] Work has probably grown more intense, because of globalization and competition, and technologies further blur boundaries between work and home. Americans have about the same amount of leisure time as before, but more choices of activities to fill that time.[30]

The added pressures help explain the systematic cultural response of more fast food, and more need for the health club to dissipate stress—but that takes time too. Many Americans respond to time pressure by multitasking—driving with a cell phone or a cup of coffee in their free hand. How often do we get involved in two or three tasks at the same time? Not long ago a valued colleague turned down an invitation to lunch because it would mean time away from an important writing task: he would eat lunch at his desk.

But television is the greatest devourer of leisure time.[31]

Television Devours Time, but So Do Computers

Researchers have studied Americans' use of time via time diaries. If we subtract sleep and eating and grooming from the 168-hour week for people ages18–64, we get roughly 100 hours a week to divide between work, family care, personal care, and free-time activities. If half of that 100 hours goes to paid work and attending to family and intimate relationships (a number roughly equal for men and women), that leaves 40 hours for free time to do whatever we like, with 10 hours allocated for personal care.[32]

Advances in technology have made housework much less onerous and time-consuming. One hundred years ago, housekeeping involved about 56 hours per week. It took 4 hours to do a load of laundry and 4.5 hours to iron it.[33] Today, once we load the dirty clothes into the washing machine and then the dryer, we are free to do other things. Do modern fabrics actually need any ironing? One hundred years ago a clean house, cooked meals, and clean clothes were a luxury, but now they are basic necessities, accomplished more quickly—or simply purchased. What do we do with all the time saved?

Evidence is clear that at least half of Americans' free time is now spent watching television,[34] averaging almost three hours a day, with some watching much more. (Researchers encountered people who said they don't have any free time *because* they were watching television.[35])

People still watch television, but within the past decade there has been a huge switch in media use among teens and 20- and 30-year-olds. This generation has shifted their attention and use of time to computers—even though they are often watching videos on their computers or cell phones.

The numbers are startling. A Kaiser Family Foundation study reveals, "Over the past five years, young people have increased the amount of time they spend consuming media by an hour and seventeen minutes daily, from 6:21 to 7:38—almost the amount of time most adults spend at work each day, except that young people use media seven days a week instead of five."[36]

Because of multitasking and using more than one medium at a time, this young generation packs "a total of 10 hours and 45 minutes worth of media content into those daily 7½ hours—an increase of almost 2¼ hours of media exposure per day over the past five years. Use of every type of media has increased over the past 10 years, with the exception of reading. In just the past *five* years, the increases range from 24 minutes a day for video games, to 27 minutes a day for computers, 38 minutes for TV content, and 47 minutes a day for music and other audio."[37]

Smart phones, iPads, iPods and other MP3 devices, high-definition screens, YouTube, social networks like Facebook—all contributed to an explosion of media use among American youth. Use of MP3 players and cell phone ownership and use (with texting) is spreading to younger teens, with the majority of teens now owning a cell phone. Texting has become the preferred channel of communication for teens, with half sending 50 or more text messages a day, and one in three sending more than a 100 messages a day.[38] Young people watch less of *regularly scheduled* television programming, but their use of television content has increased via the Internet or cell phones. Older adults continue to watch more television rather than spending time on the Internet. Indeed, some research suggests that adults over 18 years old watch 319 television minutes a day, which is double the time they spend on the Internet, 156 minutes a day.[39]

Such total media consumption, with often the use of two media at a time, feeds the modern experience of hyperstimulation, floods of information, and the acceleration of time.

Speed Is Addictive

Speed is addictive. James Gleick points out that English has a special word with a dual meaning: *rush* as a noun means exhilaration, as in that drug gives me a terrific rush. *Rush* in the more conventional meaning implies hurrying and the need to speed up.[40] We have a preference for going faster rather than slower. No one wants a new machine (computer, Xerox machine, cell-phone, auto) that will do things more slowly. Most people do not want to be in a work situation where time appears to go more slowly.

Speed is contagious.[41] Slower media, such as magazines and newspapers, imitate the fastest media, such as the Internet and television. Articles become shorter and shorter with clearer messages and less analysis. Super-brief news items get updated continuously. Everyone has 10 seconds to spare, but who has a few minutes to spare? Concern to simplify and speed up information gives the edge to the fastest and most compact media. What gets lost in this speedy information environment is context and understanding. Time pressure compels consumers of information to scan multiple channels, scooping up and filtering large amounts of data in search of what is interesting or important. But they do not necessarily remember much of what they scanned; the last bit of data pushes out from consciousness the bit of data previous to it.

Speed influences style and syntax. Many do not proofread the e-mails that they write. Traditional correspondence, once mailed off, allowed the sender to turn to other activities. Now one is expected to respond within hours. Indeed, if there is not an almost immediate reply, the expectation is that the person did not receive it.[42]

Limits to Speed?

Are there limits to speed? Performance standards in sports and finance keep edging higher and faster. In the competitive world of finance, workers seek to hit home runs, that is, get great returns and downplay steady, gradual changes. Stockbrokers face daily, if not hourly, pressure to beat the market. Internet and online trading have greatly increased the efficiency and speed of the markets. Information flows quickly across markets to millions of potential traders and investors. Are the very factors that have increased the efficiency and speed of markets also leading to the increased propensity for bubbles and malfunctions?[43]

Some Olympic athletes and officials wonder if there are limits to speed. Speed played a role in the death of a luge athlete at the 2010 Winter Olympics in Vancouver when his sled spun out of control on the superfast run at the Whistler Sliding Center. The Whistler Sliding Center was marketed as "faster, steeper and more intense than any track in history," and elicited speeds of nearly 96 miles per hour in luge, a sport in which sledders wear little protection but a skintight suit and a helmet.[44] Now the sport is asking, how fast is too fast, and what are the limits of speed?

Impatience and Restlessness

The speed of modern life with its stimulating stream of information changes us. We become accustomed to instantaneous responses to our wishes and desires. This creates expectations that everything will happen quickly. Waiting on line in a store or being caught in traffic makes me impatient. Waiting feels like I am wasting time, and time is valuable. I have interviewed store clerks who tell me how quickly customers become restless or irritated if lines seem to be moving slowly.

Most people don't like waiting on line, of course, but standing on an orderly line to wait one's proper turn once was a marker of modernity, a movement from chaos to order. In India, for example, as a growing middle class feels more well off and less survivalist, they are less eager to cut into lines or form a ragged scrum as they wait for services.[45] As scarcity turns to abundance, as impulse turns to rules, line etiquette in India becomes more common. But as time means money, impatience grows and the free market rushes in to fix the consumption of time in line. Thus, business travelers can now purchase "rush visas" that enable them to jump lines for a few hundred dollars extra.[46]

I have a sense of time theft if I am caught in a useless meeting or trapped in a situation other than one in which I could be more active and more productive. I feel this sense of, "Don't waste time! Hurry up!" I live with an urgency that I should be doing *something,* that I must use this time and not waste it. The press of time, the time imperative, time pressure, are part of the modern experience that makes many of us restless and impatient with the slow pace of many aspects of life, whether waiting for our calls to be answered or our computers to boot up or our boring meetings to end.

Neil Gabler comments on our modern impatience. Decades ago, he says, during the Great Depression, few Americans expected an immediate remedy. What they expected was some form of action. In our current economic difficulties, the public attitude is different. People want the current economic disaster to be over *now.* Says Gabler, "Unlike our forebears, we live in a society in which nearly everything happens instantly. Impatience is the new American way."[47]

Life Outside of Time: Slowing Time

Many of life's activities proceed according to their own internal time: pregnancy, love, flowers blooming, bread baking. When we strap ourselves to our watches, we signal our subordination to clock time. When we take our watches off to go on vacation, we slow time down. Attending to family and friendship involves slow time. If we get away and escape daily tasks and turn off cell phones and modern media, we enter slow time.

It appears more and more difficult to enter slow time. For contemporary Americans, time is money, valuable, something not to be wasted but spent carefully. Bodil Jonsson reminds us that the average lifespan lasts about 30,000 days; people experience a sense of having too little time, and don't feel they have control of their time.[48]

Even children's time has changed. Children used to be able to do nothing, by entertaining themselves with made-up games or spending time on aimless activities. But our culture privileges productive time, constant change, and progress. To work and be responsible are crucial to adult self-image, especially to middle-class busy individuals. Time is precious, to be filled with purposeful activity. Doing nothing, killing time, offends against the sense that time is valuable.[49] Sophisticated computer games and the Internet hold children's attention and fill time. It is no accident that children's time is now more structured and supervised, more filled with ballet lessons and soccer games. This structure is not bad, but highly illuminates how we view time.

What price do we pay for speeding along, with so many stimuli? Some critics suggest that many of these forces work against depth and self-reflection, that many ideas have no context, and that all information is equally significant.[50]

Have we lost the ability to be quiet and passive? The Internet and video games accustom players to be interactive. Television, seeking to be less passive, now regularly asks us to share our opinions, to vote who our favorite "American Idol" is. Televised sports shows have polls that solicit our opinion on who we think will win the World Series or the U.S. Open. On websites or on Amazon we are invited to "Post a Comment" or "Post a Review." The local television news wants us to send in news tips and share our pictures and comments.

Numerous reminders prompt us to seize moments outside the pace of contemporary life. Websites and books urge us to simplify our lives, to winnow down our possessions to "100 personal items."[51] Some research suggests that people are happier when they spend money on experiences rather than material objects.[52] Dozens of books offer advice on organizing our lives and reducing clutter, simplifying life to reduce stress.[53]

One of the consequences of a speeded-up culture is that we are sleeping less. The National Sleep Foundation reports that in just a decade society has become even more around-the-clock and more complex. The National Sleep Foundation tells us, "With the advent of the Internet, cell phone, BlackBerries, we're seeing our society is increasingly 24–7. People are able to be active at any time from anywhere, and it causes people to be more active around the clock. This increased activity is essentially giving sleep less importance."[54]

The mean hours of sleep on a weeknight dwindled from an average of 7 hours in 2001 to 6.7 hours in 2009, which represents a drop of about two hours per night since the 19th century, one hour per night over the past 50 years. It could be that life in 2009 has more distractions, says Dr. David Schulman, the medical director of the Emory Clinic Sleep Disorders laboratory in Atlanta, Georgia. "It's been entertainment, Internet, playing games or TV. People have all sorts of distractions they didn't have back then (in previous years)," Schulman comments. Centuries ago, people routinely slept eight to nine hours a day, he said. But now, only about a quarter in the survey reported getting eight or more hours of sleep.[55]

Contact with Nature can refresh our brains and reverse some of the effects of heavy use of digital devices and the time demands of contemporary life. Multitasking and the demands of processing so much information deplete energies and cognitive performance.[56] Walking in the woods seems to help new learning, perhaps even more than a walk on a busy urban street, which stimulates and absorbs some of our scarce attention. If nothing else, some return to nature and its rhythms helps slow time down. Activities such as jogging or yoga can clear the mind, if one's cell phone is off and, depending on personal preference, if one's iPod is left at home.

Leisure time is a different quality and state of time. Leisure activities provide direct enjoyment. Although the number of hours spent on purely enjoyable actives hasn't changed that much in the last century, the activities have. But my emphasis here is not on actions or activities but a way of thinking about time.

We have choices about the activities that contribute to the slowing down of time or standing outside of the everyday stream of events. Perhaps we can choose to walk rather than drive an auto, to carve out chunks of time during which we have no agenda, during which we do not have to be productive. To discover quality time is to rediscover leisure. Leisure, of course, is the final luxury, the freedom from the demands of time, the freedom for creativity and enjoyment. Leisure could be called the basis of culture. Underneath ways of thinking about time and change is the larger philosophical question of what is the good life. We need reminding that it is an ever-new task to find ways of thoughtfully managing our lives so that we are fully ourselves, our best selves. It means being more self-conscious, more deliberate, about the process of living and of choosing our activities. Rare and special moments of being in harmony with the flow of life reward us richly.

Some people use meditation to help slow down time by reducing stimuli and the sense of a lot of movement and change. With practice, meditation or a walk with the cell phone turned off can sometimes provide that sudden illumination that comes effortlessly and without trouble, a kind of peak experience in which we have spiritual insight into our lives and our reality. Getting away on a vacation, to a place where the stimuli are different and fewer (if we don't constantly check our e-mail and text messages), can help slow time and promote relaxation.

Some people do not have control over time. Unemployment or reduced finances or ill health lessen one's control over one's life and time hangs heavily.

In a book published in 1947, Joseph Pieper observed that culture depends on leisure. Pieper raised the question of whether it is possible to preserve leisure within the world of work. In the highly industrialized Western world, the concept of work and productivity has an almost demonic power to keep us from reflection. The deprivation of leisure amounts to a spiritual affliction in which the worker is consumed by the working process, which values him only according to his usefulness in production.[57] In this context leisure becomes another word for laziness, idleness, or sloth.

Western culture has seen many changes, especially in the rise of intellectual worker or knowledge worker. Intellectual work, of course, has been around for centuries, but very few had the time or luxury for spending their time on gaining and disseminating knowledge.[58] With the rise of the digital economy and so many people working in information industries, we seem to have lost the insight that leisure is not laziness, idleness, or boredom.

Leisure implies an attitude of nonactivity, of inward calm, of silence; it means not being busy, but letting things happen. Leisure is a form of that silence that is the prerequisite of a deeper apprehension of reality. Leisure implies a certain serenity flowing from our recognition of the mysteries of the universe.[59] Do modern people lack the capacity to enjoy leisure in the sense of that special receptive kind of contemplation? Are we so out of training for it, out of practice, because we are never off?

7

Families, Women, and Sex

Families Are Different Now

Marriage and family life have, to quote author Paul R. Amato, "changed radically" in the last two or three decades.[1] Our experiences and perceptions tell us this, and data confirms it.

Throughout history, marriage has provided structure and meaning for men and women. It has been a marker for leaving home, forming one's own family, initiating sex, and having children. Husband and wife were roles that guided daily life and contributed to one's identity.[2]

Now, multiple forces in society and the economy are causing relationships between the sexes to shift, with profound social consequences. Families are getting smaller, cohabitation has vastly increased, and divorces are stabilizing after decades of increasing. A huge number of children are born outside of marriage, and parenting styles are changing. Adolescence is extending further into adulthood, and people are living longer than ever before.

Families are getting smaller. Households have changed over the past century, decreasing by two bodies over the past century. In 1900 the average household size, according to U.S. Census figures, was 4.60 people, shrinking to 3.68 in 1940 and to 2.61 in 2008.[3] Some of these changes can be attributed to the general increase in wealth resulting in fewer extended families since, for instance, grandparents can afford to live separately from their adult children. More affluent families generally choose to have fewer children. The technology of birth control provides the choice of becoming a parent rather than resigning to pregnancy as an inevitability. No previous generation has been so free to choose having children or not. Further, the very idea of parenthood has expanded in different ways through technologies such as in vitro fertilization that enable people to become parents who previously would have been unable.[4]

Living Together

Cohabitation before marriage, once rare, is now the norm according to the most recent statistics from the National Center for Health Statistics, the principle government source for data relating to the health and well-being of the American population. The majority of people who now get married live together before the wedding.[5]

In 2009 in the United States, 67.5 million opposite-sex couples lived together; 60.8 million were married, and 6.7 million were not married but living together. That translates into approximately 9 percent of the general U.S. population of men and women ages 15–44 who are cohabitating.[6] More than half of those cohabiting for the first time will transition to marriage within three years.

Forty-six percent of American women aged 15–44 are currently married, and 42 percent of men the same age are married. Men and women with more education are much more likely to be married than those with less education. Seventeen percent of women aged 22–44 who did not have a high school diploma were currently cohabiting, compared with 5 percent of women with bachelor's degrees or higher. Women aged 22–44 with bachelor's degrees or higher were more likely to be currently married (63%) than those who did not have a high school diploma (49%). Comparable percentages for men were 62 percent and 53 percent, respectively.[7]

All of these figures confirm that union formation, whether by marriage or cohabitation, is one of the primary events in adulthood. Indeed, cohabitation is increasingly becoming the first important union formed among young adults. From1987 to 2002, the percentages of women aged 35–39 who had ever cohabited doubled, from 30 percent to 61 percent. Over half of marriages from 1990 to 1994 among women aged 19–44 began as a cohabitation.

Some couples view cohabitation as a step in the courtship process, falling somewhere between steady dating and marriage. Many of these couples use the period of cohabitation to assess their compatibility for marriage. Other couples see cohabitation as a convenient relationship—a union that provides economic benefits (household economies of scale) combined with the availability of a regular sexual partner.[8]

Unmarried Women Giving Birth

The Center for Disease Control's National Center for Health Statistics reports a total of 4,265,555 births in the United States in 2006. A whopping 38.5 percent of those births were to unmarried mothers. This figure of more than 1.6 million babies born to unmarried women, more than one-third of all births, is the highest number ever recorded in the United States.[9]

The number of these births to unmarried women has been increasing each year, 8 percent greater than in 2005 and a 20 percent increase from 2002. In 1970, the number of children born outside of marriage was about 10 percent of all children. "All measures of unmarried childbearing reached record levels in 2006," claims the NCHS report.

The majority of all nonmarital births (52%) occurred to women cohabiting with a partner. Consequently young children are also more likely than in the past to live in a such a cohabiting household. In 2002, 2.9 million children under age 15 lived with an unmarried parent and his or her unmarried partner. Estimates suggest that about two-fifths of all children will spend some time in a cohabiting household before age 16 years.[10]

The sharp rise in nonmarital birth rates for adult women (20 and older) in combination with comparatively smaller increases among teenagers (10–19) has resulted in a continued shift in the age distribution of unmarried mothers. Whereas 4 in 10 nonmarital births were to teenagers in 1980, by 2006, this fraction dropped by nearly one-half, to just over 2 in 10. The number of births, the birth rate, and the percentage of births to unmarried women are all important indicators of childbearing patterns and changes in family formation.

If there is any good news in these figures, it is that in 2006, 73.6 percent of women who gave birth had at least a high school diploma or higher and 23.3 percent had a bachelor's degree or higher. The educational attainment of women giving birth has risen substantially over the last few decades; the increase has slowed somewhat over the last decade, however. This trend in part reflects increases in educational attainment of all women during this time. Maternal education plays a profound role on the number of births and the risk level of births. Women with higher educational attainment are more likely to desire and give birth to fewer children, and are less likely to engage in behaviors detrimental to health and pregnancy.

Teenage pregnancy and childbearing are of considerable concern because of risks to the babies at birth and long-term. The children of teenage mothers are more likely to have lower cognitive attainment at kindergarten entry, exhibit behavior problems, have chronic medical conditions, and rely more heavily on publicly provided health care; they are also more likely to be incarcerated at some time during adolescence, drop out of high school, give birth as a teenager, and be unemployed as a young adult. Compared to women who delay childbearing until the age of 20 to 21 years, the teenage mothers themselves are more likely to drop out of high school and to be and remain single parents.[11] A recent study found that the public costs of teenage childbearing in the United States are about $9.1 billion annually. This figure includes costs for various antipoverty and school programs, as well as lower levels of taxes paid by individuals whose earning potential has been compromised by growing up in poverty.

The number of births, the birth rate, and the percentage of births to unmarried women are all important indicators of childbearing patterns and changes in family formation. The number of children born outside of marriage has increased steadily, from 11 percent of all births in 1970, to more than one-third of all births today. The combination of sharply rising birth rates for *unmarried* women together with relatively stable rates for *married* women has resulted in continued increases in the *proportion* of births that are to unmarried women. Like the number of births and the birth rate, the proportion changed relatively little during the years 1998–2002, but has since climbed sharply, reaching 38.5 percent compared with 34.0 in 2002. The overwhelming majority of teenage births has long been to unmarried women (rising from two-thirds in 1990 to 84 percent in 2006).[12]

The percentage of children born to unmarried mothers varies considerably by race and ethnicity. For example, data indicate that the percentage of nonmarital births was 16 percent among Asian Americans,

31 percent among non-Hispanic whites, 46 percent among Hispanics, and 69 percent among non-Hispanic blacks. It is likely that economic as well as cultural factors account for these variations. Half of nonmarital births occur to cohabiting parents. Most of these couples view marriage favorably, however, and most claim that they are likely to marry. For many unmarried parents maintaining a relationship requires overcoming a variety of obstacles, such as poverty and unemployment.[13]

Where Are the Fathers?

About one-half of previously married cohabiters and about one-third of never-married cohabiters have children living in the household. In most cases, these are the children of only one partner. Hence, these families are structurally similar to stepfamilies. Nevertheless, a substantial proportion of nonmarital births (40 to 50%) occur within cohabiting unions. In these cases, children live with both biological parents. But because these unions tend to be unstable, the majority end in "informal divorces." Most children born to cohabiting parents consequently will spend time in single-parent households.[14]

Where are the fathers of these children? About three-fourths of fathers (73%) lived with all of their children; 14 percent did not live with any of their children, and 12 percent lived with some but not others. Of single-parent households, in 2009, 1.7 million were father-only family groups with children under 18, while mother-only family groups numbered 9.9 million.[15]

The structure of families diverges drastically by social class. Among women with no more than a high school education, the out-of-wedlock birthrate has grown rapidly since the 1960s and is now approaching half of all births. In contrast, single motherhood is still rare among college graduates, representing less than 5 percent of births among this group overall. Almost all college graduates still marry eventually, but marriage rates are dropping steadily among those without a high school degree. Divorce has declined among the well-off since the 1980s but is climbing among the unskilled. Although white marriages have achieved greater stability over the past 30 years, black marriages at every level dissolve more frequently. As a result, many more black children in all income brackets grow up with one parent.

The offspring of the well-off receive a growing share of parental time, attention, and investment and grow up in stable and orderly homes. The less privileged frequently endure a fractured and chaotic family life.

It is not news that America is a land of the haves and have-nots. But it is nevertheless troubling to see once again how in many, many ways the rich get richer and the have-nots get poorer and poorer. America is increasingly a country of those who are making it and those who are having a harder and harder time being successful.

But even intact, apparently successful families are under stress.

A major preoccupation for many busy parents is the juggling of family time. "You take Sally to ballet, and I'll pick up Nick. And get some pizza on your way home, since I have a school meeting tonight." Grazing, snacking instead of sitting down for leisurely meals, and watching television during mealtimes instead of conversing have undermined the shared meal and thereby damaged family life. Author Michael Pollan comments on "the family meal—where children learn the art of conversation and acquire the habits of civility—sharing, listening, taking turns, navigating differences, arguing without offending—and it is these habits that are lost when we eat alone and on the run."[16] Parents necessarily hand over to others (schools, coaches) functions that they might have performed for their children in generations past. "The faster and more encompassing changes are, the more problematic the transition of culture between the generations becomes. Children and adolescents are increasingly free to fashion their own values and their own meaningful lives, pulling together fragments from various sources," remarks T. H. Eriksen.[17]

The Economic Rise of Women and the Eclipse of Men

Startling shifts in society emerge into view as we look closely at the rapidly shifting landscape of the sexes. Two key shaping trends emerge: economics and education.

Women entered the work force during world War II and never left. The vast majority of women are in the work force, and their involvement in paid work has reached a historical high point. Women now outnumber men in the work force. This tipping point was reached because of high male unemployment during the Great Recession of 2008 and 2009 when men lost three-quarters of lost jobs, that is, men accounted for 6 of 8 million lost jobs. Industries that lost most jobs tend to be cyclical and overwhelmingly male, such as construction, manufacturing, and finances.[18] About 60 percent of American women are now in the work force, up from 43 percent forty years ago. In that same period the proportion of American men in the work force has declined from 80 percent to about 73 percent.[19]

Of the 15 job categories projected to grow the most in next decade, all but two are occupied primarily by women. Men dominate just two of the 15 job categories projected to grow over the next decade: janitor and computer engineer. Women have everything else: nursing, home health assistant, child care, food prep, and so forth. The postindustrial economy is indifferent to men's size and strength. The attributes that are most valuable today (social intelligence, communication skills, the ability to sit still and focus) are at a minimum more common in women and not usually thought to be male strengths.[20]

Men still dominate the upper reaches of society, but given the trends pushing the economy and society, this domination seems like a last gasp of a dying age. The role reversal that's under way between American men and women shows up most obviously and painfully in the working class. The Rust Belt and other places of our postindustrial society have turned traditional family roles upside down. Manufacturing has declined, with almost 6 million jobs lost since 2000, more than the total of its work force. In 1950, 1 in 20 men of prime working age was not working; today the figure is about 1 in 5—the highest ever recorded. This economic and cultural power shift from men to women would be hugely significant even if it never extended beyond the working class. The transition is moving up into the middle class. Women are starting to dominate middle management and a surprising number of professional careers. Women now hold half or more of the positions in managerial and professional jobs, up from 26.1% just 30 years ago. Women hold half or more of accounting, banking, and insurance jobs. About one-third of all physicians are currently women, as are almost half (45%) of the associates in law firms. A white-collar economy values raw intellectual horsepower, which men and women have in equal amounts. It also requires communication skills and social intelligence, areas in which women, according to many studies, have a slight edge. Also increasingly the modern economy requires formal education credentials, which women are more likely to acquire. The only professional areas where women still make up a relatively small minority are engineering and those calling for a science background—but even here women are making strong gains. It is only at the very top of the jobs pyramid that the upward march of women stalls. Only 3 percent of the chief executive officers of Fortune 500 companies are women.[21]

Women at the top of their profession seem to pay a higher price than men do, specifically in regard to marriage and children. The last three men to be appointed to the U.S. Supreme Court all were married with children. The last three women who were appointed (Elena Kagan, Sonia Sotomayor, and Harriet Miers—Miers withdrew before being confirmed) have all been single and without children. The pay gap between men and women is still not zero, although it has shrunk to just a few percentage points.[22] Many more women take time off from work or are working part time at some point in their careers. Our economy exacts a steep price for any time away from work in both pay and promotions, with a harsh price for pursuing anything other than the old-fashioned career part. Women do almost as well as men today, as long as they don't have children. A disproportionate number of prominent women do not have children, including for example former U.S. Secretary of State Condoleezza Rice and current U.S. Secretary of Homeland Security Janet Napolitano.

Women are also a growing presence in arenas where they have largely been absent before now. Title IX is one of the key areas of a 1972 law against discrimination in 10 key areas. Most people think of it primarily in regard to sports.[23] The value of the law was to grant females equal access to sports. In the academic year 1971–1972 less than 300,000 females were involved in high school sports. In 2007–2008 that figure had increased 940 percent to more than 3 million. On the college level participation went from fewer than 30,000 in 1971–1972 to more than 160,000 in 2004–2005, a 456 percent increase.[24] There are still inequities in sports, specifically in money and scholarships and even participation, but women's presence is much more in evidence—whether it's in the Olympics or in nontraditional sports like full-tackle football, hockey, and wrestling.[25]

Educational Differences

If the economic numbers are surprising, the education figures are even more striking, giving a glimpse of the future. Educationally men are on a down escalator and women are on a rapid up escalator. Census figures are startling. Far more women than men are expected to occupy professions such as doctors, lawyers, and college professors since they represent approximately 58 percent of young adults, age 25 to 29, who hold an advanced degree. In addition, among all adults 25 and older, more women than men had high school diplomas and bachelor's degrees.

Among people in the 25–29 age group, 9 percent of women and 6 percent of men held a master's, professional (such as law or medical), or doctoral degree.[26] Needless to say, there is an economic advantage to holding advanced degrees. Even though women still lag behind men in earnings, the future directions are very, very clear.

Fifty years ago the percentages of advanced degree holders by gender were entirely different: in 1960 men held 78 percent of advanced degrees, women 22 percent. In 2009 women earned 58 percent of advanced degrees, versus 42 percent earned by men. At the bachelor's degree level, 35 percent of women and 27 percent of men held this degree, which is the entry passport to the modern white-collar economy. This gap has grown considerably in the last decade: it was only 3 percentage points in 1999 (30% for women, 27% for men).

There are social consequence of this growing education gap. As the pool of educated professional women grows, the pool of their male counterparts shrinks. Many educated, well-paid women may well have to settle for a romantic or marriage partner who is not her equal economically and educationally.

If we look at boys and girls still in school, disparities continue. Boys continue to amble along while girls are sprinting. Increasingly colleges provide affirmative action with boys playing the role of the underprivileged applicants needing an extra boost. Colleges seek an appropriate gender balance, a tipping point, where 60 percent or more of their enrolled students are female. "The reality is that because [academically well qualified] young men are rarer, they're more valued applicants. Today, two-thirds of colleges and universities report that they get more female than male applicants, [and] . . . that the standards for admission to today's most selective colleges are stiffer for women than men."[27] Schools, like the economy, value the self-control, focus, and verbal aptitude that seem to come more easily to young girls. It is terrific to see girls and young women poised for success in the coming years. But allowing a generation of boys to grow up feeling rootless and obsolete is not a recipe for a healthy, peaceful society. Far from being celebrated, as it should be, women's rising power is perceived as a threat.[28] It should be noted that among students from the highest income families, the gender gap among college students largely disappears.

Changing Views on Marriage

Why so many changes in basic institutions like marriage and family?

How people think about marriage and relationships has changed. Some experts theorize that a growing individualism in American culture is preoccupied with the pursuit of personal happiness such that people are less willing to be hampered with obligations to others. Thus commitments to traditional institutions, which require social and moral obligations, will erode. If this is valid, a different notion of marriage may be emerging, one in which marriage is a path toward self-fulfillment, a voluntary relationship in which people have high expectations for personal satisfaction. This view is far from traditional marriage, which was an institution primarily for economic security and procreation. The traditional view constrains one's behavior and emphasizes obligations to others. Several consequences flow from this shift in our way of thinking about marriage: the increase in nonmarital cohabitation, the relative ease of divorce, less stigma regarding children out of wedlock, and more diversity in marriage types, such as gay marriage and blended families.[29]

Those Who Do Marry Have Many Choices as to Kinds of Weddings

If single people are having difficulty finding a suitable romantic partner, an abundance of online dating services is glad to help them find a suitable match. There are at least 900 dating services, such as the online sites eHarmony.com and Match.com, which earn approximately half a billion dollars for their assistance.[30] Sites less oriented toward marriage cater to those singles who prefer short-lived pleasures over permanence, such as Lavalife.com.[31] Couples who meet and do want to have a wedding can choose from a greater diversity of wedding types, as well as prewedding and postwedding events. One fairly recent and evolving custom is the bachelorette party, probably modeled on the bachelor party. The bachelor party is probably centuries old, while the bachelorette party may have begun during the sexual revolution of the 1960s, but remained uncommon until at least the mid-1980s. The bachelorette party reflects the increase of gender equality and the economic independence and clout of women. As to the wedding ceremony itself, couples increasingly choose to customize and personalize their commitment ceremony.[32]

Currently one can choose a destination wedding, a cruise wedding, a beach wedding, and so forth. Weddings can be in costume and push the edge of conventionality. YouTube documents the extent to which some couples seek something unusual.[33] One can have a friend or member of the family perform the wedding merely by applying to the appropriate state office. Some states have laws that permit persons (other than a rabbi, priest or minister, or judge) to apply for authority to perform marriage ceremonies. The grant of authority is valid for one day—and thus a lay person may officiate at the wedding of family or friends on that one day.[34]

In 2009 there were about 2,152,000 weddings. It is difficult to calculate the average spent on a wedding since the wedding industry is made up of multiple smaller enterprises like caterers, wedding consultants, dresses, beauty suppliers (hair, makeup), photographers, gifts, music, travel, and honeymoon. That said, an average wedding may cost about $20,000, but the wedding industry as a whole represents a lot of money, about $42 billion.[35]

Gay Marriage

Gay marriage simply wasn't in public consciousness a generation ago. Now gay marriage is very much a contentious issue for many people, with much of the debate surrounding the changing concept of the institution of marriage. Marriage as we know it today has only existed in its current form for the majority of the 20th century; marriage is an evolving and constantly changing construct, and has served many social purposes. Marriages in many Western and Eastern cultures have traditionally been property transactions or arrangements between families. Currently some states allow marriage between same-sex partners, some states allow civil unions, and some states continue to have legal marriages only between opposite-sex partners.[36]

Divorce

Marriage responds to currents within the culture and hence has always been in flux, perhaps more so than we recognize. If the change in how people view marriage has increased people's expectations for personal happiness, it is not surprising that divorce rates have increased during the second half of the 20th century. Those who want can leave an unsatisfying spouse and find another romantic partner.[37] The divorce rate has been increasing gradually, in general, throughout American history. The rise during the 1970s, however, was particularly dramatic, with the rate doubling in a single decade. Since reaching a peak in the early 1980s, the rate appears to have declined. The crude divorce rate (defined as the number of divorces per 1,000 population) rose from 2.2 in 1960 to a high of 5.3 in 1981 and then declined to 3.7 in 2006. These figures suggest a 28 percent decline in the divorce rate since 1981.[38]

Divorce may be good for unhappy adults who flee one partner to find happiness with a new spouse, but divorce is generally not good for children. Nearly one million children experience divorce each year, and about 40 percent of all children with married parents will experience divorce before leaving their teens. The high rate of marital disruption puts about half of all children temporarily into single-parent households, usually with their mothers. Married couples with children enjoy, on average, a higher standard of living and greater economic security than do single-parent families with children. Thus, the child poverty rate is more than four times higher in single-parent households than in married-couple households The economic advantages of married couples are apparent across virtually all racial and ethnic groups. Research consistently shows that children who experience divorce, compared with children who grow up with two continuously married parents, are more at risk of conduct disorders, psychological problems, low self-esteem, academic failure, and difficulties forming relationships.[39]

Possible Declines in Intimacy and Communication Skills

Are people's interpersonal skills eroding? Has the pace of life and increased reliance on communication technologies rendered people less capable of face-to-face intimacy and communication? Some researchers find evidence of noticeable changes.

Americans have fewer close ties to those from their neighborhoods and from voluntary associations. Sociologists suggest that new technologies, such as the Internet and mobile phones, may play a role in advancing this trend. Communication technologies support relationships that are relatively weak and geographically dispersed, not the strong, local ties that tend to be a part of peoples' core support.[40] They depicted the rise of Internet and mobile phones as one of the major trends that pulls people away from traditional social settings, neighborhoods, voluntary associations, and public spaces that have been associated with large and diverse core networks.

Among adults, husbands and wives report to researchers that they have fewer close friends and people to confide in. Further, spouses say they have fewer shared close friends, and a growing number confess that they have no close friends at all. Close relationships appear to have dropped in number, with people having fewer contacts from clubs and neighbors. Americans' dependence on family went up from 57 percent in 1987 to 80 percent in 2004.[41] Affiliations with community organizations have declined. In short, there appears to be a marked decrease in social participation, although ties to religious organizations have grown. Married couples increasingly eat alone, and interactions in general have declined.[42]

Among college students? Some college officials wonder if overly protective parents reduced the ability of teens and young adults to work out everyday social problems with roommates. College officials who work with young people raise concerns about students lacking the skills or the will to address ordinary conflicts. This lack is noticeable in college dorms. Says Tom Kane, director of housing at Appalachian State University in Boone, North Carolina: "We have students who are mad at each other and they text each other in the same room. So many of our roommate conflicts are because kids don't know how to negotiate a problem."[43] Another director of housing at the University of Florida, Norbert Dunkel, says, "It used to be: 'Let's sit down and talk about it.' Over the past five years, roommate conflicts have intensified. The students don't know how to have the person-to-person discussions and they don't know how to handle them."[44]

Could it be that reliance on cell phones and Facebook to communicate have made it easier for young people to avoid uncomfortable encounters? Surfing the Internet is a solitary activity, done alone, even though one might connect via social networks with others. Possibly smaller families may have produced many teens who never had to share a bedroom with a sibling and thus did not learn negotiating skills in intimate situations—borrowing clothes, keeping a room picked up, and so on. Sarah English, director of housing and resident life at Marist College Poughkeepsie, New York, says, "It surprises me when students say, 'My roommate's mother called me and yelled at me.' "[45]

Students' cell phone connection with parents enables protective parents to jump in quickly. In an age of Skype and twice-daily texts home, colleges are trying to encourage "Velcro parents" to back off so students can grow up and develop independence.[46] Some college officials administrators see parental overinvolvement in ways that were not possible in previous years.

Isn't it easier to vent in a text than express anger directly to a person? A California high school history teacher raises the question. As her students are increasingly immersed in texting, Nini Halkett finds them increasingly shy and awkward in person. "They can get up the courage to ask you for [a deadline] extension on the computer," she says. "But they won't come and speak to you face-to-face about it. And that worries me, in terms of their ability—particularly once they get out in the workplace—to interact with people."[47]

Robert Putnam raised these issues a decade ago.[48] Even if the trend is overstated, enough evidence exists to say that personal relations have not so much eroded as evolved in ways that researchers haven't quite captured. Women are working more, so perhaps social ties at work have become relatively more important. Perhaps the nature of civic participation has changed. The conclusion seems to be "that social connectedness has *changed* rather than *declined*."[49]

Extended Adolescence or Emerging Adulthood

Being parents in a complex, changing society is difficult, and most parents need to rely on institutions outside the family to assist in socializing and training their children for entrance into society. In a less complex era, families were more self-sufficient and better able to provide most of what their children needed.

"Emerging adulthood" seems to be recognized as a distinct developmental period of the life course for young people in postindustrialized societies. During this period of age 18–25, young people undergo psychological changes, explore opportunities, and move toward enduring choices of love, work, and worldviews. Not everyone goes through this period at the end of the teen years, but it seems to be more common in modern societies, which require high levels of education and training for entrance into information-based professions. Many young people remain in school into their middle twenties and early thirties. They typically postpone marriage and parenthood until after their training has ended, which allows for a period of exploration of relationships outside marriage and experimenting with various jobs before taking on the responsibilities of parenthood.[50] This stage of life allows the freedom and time to explore many different futures as possibilities. It is also a period of life that is likely to grow in importance in the coming decades, as countries around the world reach a point in their economic development where independent adulthood requires a delay in taking on responsibilities and affords the possibility for exploration.[51]

Other Developmental Changes

No matter how complex the society, good medical care and information enable people to live more healthy lives and longer lives. A century ago, life expectancy at birth was 47.3 years. In 2006, life expectancy at birth was 77.7 years for all races. White females continue to have the highest life expectancy at birth (80.6 years), followed by black females (76.5 years), white males (75.7 years), and black males (69.7 years).[52] It helps to be well educated and reasonably affluent. Of the MIT Class of 1950 a remarkable 63 percent of the 1,030 graduates are still alive, 646 men aged 80 or more who walked at their graduation with survivors of the Class of 1900.[53] Part of my research on this book has been to interview several individuals in their eighties and nineties—one of whom said, "I never expected to live this long. Okay? Now, are you ready for another set of tennis?"

8

Making Sense of Contradictory Social Trends

Who Are We? What Have We Become?

Talking with friends and relatives and colleagues makes clear how much I have changed. Could it be just me or have my surroundings and the American culture also greatly changed?

Preceding chapters have highlighted areas of modern change: the enormous flood of information, the high levels of stimulation from a range of extraordinary communication devices, the emergence of adaptive new behaviors, the increased pace at every level of life, the continued shift of work from industrial to information services, and the changes in families and relationships. In addition, in the last two decades Americans have fought two wars in Iraq, engaged in a lengthy war in Afghanistan, and have experienced a decline in their economy, with 14 percent of the population living in poverty (over 43 million Americans).[1]

Previous decades and centuries, of course, have also witnessed profound transitions. The industrial innovations of the Second Industrial Revolution changed lives at the end of the 19th century. The Great Depression of the 1930s, and the social changes after World War II and during the 1960s, stirred society in the middle of the 20th century. What is different now are the many changes occurring simultaneously and at great speed. Further, the sweep of change touches unexpected areas, but we have almost immediate awareness of them through modern communications. These unique technologies both spotlight and hasten the transformations.

Significant political and economic problems contribute to a complex culture in flux—or undergoing a nervous breakdown. Contradictory trends and forces puzzle and confuse people. In truth, there is a furious but indirect national debate going on about values, who we are as a country, and where we are going. Many people are fearful, many angry. But is anyone listening to anyone else? Are solutions emerging?

The following sections suggest six trends or contradictory forces in collision. Sometimes there are tradeoffs, and other times there is paralysis or denial and abandonment of the problems.

First Major Trend: Decline in Social Capital, but an Urge to Belong

Interactive

Research suggests that people do not engage in as many face-to-face encounters as they used to. People, however, seem to seek to be connected in new ways. Indeed there seems to be an urgency or anxiety to connect, to belong, be a part of something, to be engaged.

Most media—television, radio, the Internet—invite interaction. Websites and television programs and radio stations seek to have their audience participate rather than be merely passive. Thus it is common to hear from these medias, "Send us your opinion," "Let us know what you think," "Contact us with breaking news," "Become a member and get your password here," or "Your opinion is important to us; take our survey now." Viewers can vote for their choice on *American Idol* or their favorite dancer on *Dancing with the Stars*. For more than a decade, the increased numbers of talk-radio shows have engaged listeners on the air with questions and opinions.

Institutions and businesses seek to have their customers be engaged and urge them to "manage your account" online. *National Geographic* magazine invites subscribers on an insert to one of its issues, "Manage Your account 24 hours a day/7 days a week: change your address, check your expiration date, report a missing issue."[2] The benefit, of course, to consumers is increased speed and efficiency; for the business or institution, the benefit is updated information about the subscriber. Many merchants and organizations, for a variety of economic reasons, offer membership cards, reward cards and loyalty programs, which provide instant and detailed information about customers and their buying habits in exchange for sales and discounts.

Many sports fans display their membership in Red Sox Nation or display their loyalty by wearing shirts with their favorite team's logo or athlete-hero's jersey number. Some programs show behind-the-scenes photos and video clips to foster a feeling among the viewers that they are part of things, belong to a community—especially if they can visit forums or read blogs connected to a particular program or site.

The implication is that "I belong to something, I am part of this group." Branding in clothing and sports apparel cloaks those wearing these brands with a special aura: I belong or am somebody, I am among those who share in the lifestyle or image created by ads for Ralph Lauren or Izod or Aeropostale.

Digital Communities

The Internet, bestowing miracles of access and convenience, is filled with bloggers. Several websites that track blogs estimate there are more than 140 million blogs—not counting 72 million Chinese blogs—with perhaps more than 100,000 new blogs going online each day.[3] Technorati.com, which tracks blogs, became overwhelmed by the millions of blogs in the blogosphere and so has narrowed the number of blogs and categories that it tracks to about one million.

The Web and a variety of Internet sites and gamers may constitute a pseudo community, an "as if" group. We can distinguish between a *group* and a *community* if we view a group as a collection of people who may be in the same place, literally and figuratively, and who share some limited characteristics, such as common interest in a topic. A community is different in that its members share common concerns about what they hold in common, and may have closer emotional bonds. A pseudo community is something that appears to its members to be a community, but is actually a much looser collection of people whose loyalty and caring about each other is closer to the group level than that of a community.[4]

Television can be powerful in creating certain kinds of communities. Oprah Winfrey is remarkably charismatic and her TV program inspires extraordinary loyalty and enthusiasm among her fans. Oprah's first show of her final season, on September 13, 2010, showed a video clip of a fan testifying how she had watched Oprah faithfully for 20 years and calculated that she had spent 5,500 hours of watching the show. The camera panned over the studio audience, which manifested hysterical emotion, joy, enthusiasm, and plenty of tears.

An instance of many people joined emotionally in a pseudo community was the worldwide sense of loss of Princes Diana. Television offered scenes of people weeping as if they knew Diana personally. The 9/11 attacks, also, brought a sense of unity to America because of the shared experience of watching the horrors play out live on television screens. An earlier experience of America feeling as if it were a community in some way was watching television after JFK's assassination. It is quite legitimate to seek social solidarity in a time when there are fewer public ceremonies that foster bonds between diverse people. Presidential inaugurations and state-of-the-union addresses scarcely attain this standard.

The locus of one's community often has shifted to an "as-if" community, as if I knew these people who share a common interest with me. As-if or pseudo communities provide comfort in knowing there are others like me. These communities contrast with a geography-based community with whom I have face-to-face dealings, whether in a traditional bowling league or project team at work or a local church group.

Second Major Trend: Disconnected Information and Images Versus Embracing Symbols of Identity

Closely related to the yearning to belong is the need to bolster the sense of one's own identity. Media and communications technologies make the world powerfully present in the form of images, products, and ideas. With a flood of disconnected information and images streaming toward me, who am I? Who are we?

Opinion writer James Poniewozik suggests that "we are moving from the era of mass culture to the era of individual culture."[5] The monolithic mainstream culture in the mid-20th century helped define what it meant to be an American, even as it was un-American at heart because of its exclusion of diversity and uniqueness. Now we seem to be in a period of cultural atomization and fragmentation as commercial tastemakers enable an increased privatization and an a la carte culture.

For a giant diverse country to have a common, mainstream culture was impossible prior to technologies of mass communication, like film and radio and national magazines. For a period, decades ago, when there were just three national networks, millions of Americans listened to the same pop hits and watched the same television programs because there was not much else available. The mass media defined and shaped mainstream culture. Now sizable chunks of the audience, especially young viewers, can demand programming that much of society considers objectionable—the unusual or gross, or risqué or adolescent—because multiple cable channels and unlimited websites can serve it up.

Indeed, some of the previous gatekeepers of media decency find the going is much more difficult now. The Parents Television Council used to have enough clout to pressure the Federal Communications Commission to crack down on racy programs. For a variety of reasons, the organization is defanged. The entertainment industry used to be afraid of the council's wrath to the degree that Fox blurred the naked behind of a cartoon character; the industry is now pushing the boundaries of taste with renewed intensity, especially with regard to "fleeting expletives" and sexualizing of young females that "borders on pedophilia," according to the council.[6]

People with local customs and accustomed ways of doing things feel that powerful images and commercial messages can erode or threaten the familiar and a sense of who they are. Uncontrollable changes in one's world, work, and relationships contribute to anxiety and anger—even if one does not know where or how to direct the anger. To feel disenfranchised from one's own culture, to be confronted by the new or unfamiliar, can mobilize intense feelings. For many people, almost everything is different from the time of their fathers.[7]

An intense and inspiring cause sows the seed for smaller but more homogeneous groups to spring up. Religious groups that offer a greater sense of participation and belonging than the mainline churches attract converts. Tea Party activists mobilize supporters out of passion about immigrants or taxes or other emotionally charged issues. Ideologically fervent Republicans disdainfully call their less passionate party members RINOs, Republicans in Name Only. Website and video images prompt strong emotional responses as people seek to regain control over their lives and destinies, which make them feel somehow adrift. Many social and economic forces prompt the gathering together of like-minded people that hope to feel the security of being surrounded by familiar values and people who look like themselves.

Partisan newspapers used to be the norm a century ago. Currently America's news media appear to be lurching backward toward segmentation and one-sidedness that characterized the newspapers of the past. Fox News has attracted attention because of how it filters and selects what it puts on the air. Journalist John Harwood writes, "Press critics worry that the rise of media polarization threatens the foundation of credible, common information that American politics needs to thrive."[8] But decade-long trends in identity-media deepen partisan passions and concentrate and amplify strong, sometimes extreme opinions.

Social change is troubling and disruptive. Transitions ignite disputes on fundamental issues of values and basic assumptions about how the world is constructed. Culture wars erupt over different conceptions of moral authority, ideas and beliefs about truth and goodness, obligations to one another, and the nature of community. Often conflict is expressed as a clash "over the meaning of America, who we have been in the past, who we are now, and perhaps most important, who we, as a nation, will aspire to become."[9]

Author David Lowenthal posits, "Remembering the past is crucial for our sense of identity."[10] Cultural amnesia can be very troubling and corrosive of one's sense of self and meaning. Unfortunately, intellectual habits are slipping as knowledge becomes more specialized and niche-oriented. Digital media, online collectivism and an increasingly democratized marketplace challenge professionals and traditional expertise. Some observers have concerns that there is a decline of general knowledge among large segments of the population.[11] Cultural literacy refers to the idea that citizens in a democracy should possess a common body of knowledge that allows them to thrive in the modern world. Included in this body of knowledge would be such terms as dates (1776), historical persons (John Brown, Hitler), terms from science (DNA), titles of historical documents (Declaration of Independence), and so on. E. D. Hirsch, who coined the term *cultural literacy,* believes there is a decline in this shared knowledge, a decline that started even before the ascendency of the Internet.[12]

Other commentators also mourn the loss of a sophisticated literacy. Pop culture, a culture of advertisement and cynical politics, has eroded habits of mind that can examine and sort out evidence. Americans need to be equipped to analyze and critique those areas of popular culture that trivialize and vulgarize American life. Professor and commentator Alvin Kernan writes, "Americans are a different people than they once were."[13]

Where is the larger conversation about these issues? Public figures have access to communication tools and give expression to what they think is going on and what the stakes are in public culture and larger systems of meaning. These elites, by their prominent positions, create the concepts, supply the language, and explicate the logic of public discussion. Politics, of course, is the arena where the conflict gets thrashed out. Much of public culture involves competing values and ideals. At the heart of public culture, then, is religion in the broad sense of faith and belief that provide our most deeply held ideals of right and wrong, good and bad, the social and moral meaning of marriage and the family.[14] And it is in this realm of belief where large doses of emotion and the nonrational (not necessarily irrational) can emerge. Public arguments commonly are based more on feeling than on fact and evidence.

Unfortunately, the quality of public discussions seems to have declined in our "post-truth era."[15]

Third Major Trend: The Blurring of Fact and Fiction

Truth has become more slippery in public discourse. Evidence-based facts are harder to find. On the political stage unfounded accusations can take on a life of their own when they get magnified by transmission across the Internet.

Examples of the debasement of the nation's speech can be found in all the media where there is a continuous blurring of clarity, evidence-based fact, and intellectual discrimination. The main political parties used to kept the extremes of right and left somewhat under control. The information age has speeded up the decentralization of political communication.

The fragmentation, enabled by the Internet, makes it hard to find the center. Today's media—especially the Internet—possess the power to amplify and spread error with an efficiency that astonishes.[16] Recent political campaigns and Congressional battles prompt a rapid rise in stubborn but false rumors or political attacks. Once a conspiracy theory or false accusation has gone viral, it seems impossible to slay these beasts in spite of an avalanche of evidence laying out what is false.

Despite all the evidence, many still believe President Obama is a Muslim and wasn't born in the United States. Maybe the truth isn't what it used to be. A CNN-Opinion poll found that 27 percent of Americans believe the president was not born in this country.[17]

Another example of this kind of unsubstantiated attack is an ad from a group of former Swift Boat veterans that claims presidential candidate John Kerry lied to get his decorations for bravery. None of those in the attack ad by the Swift Boat group actually served on Kerry's boat. And the veterans who accused Kerry were contradicted by Kerry's former crewmen and by Navy records. Nonetheless, "to swift-boat" someone has entered into the political vocabulary for spurious political attacks without a basis in reality.[18]

An example of a similar kind of viral cyber-conspiracy theory, spread by partisan bloggers, anonymous e-mailers and talk radio, occurred during the health-care bill's Congressional passage during 2010. "Death panels" was a phrase used with the assertion that the government would set up boards to determine whether seniors and the disabled were worthy of care—spread through newscasts, talk shows, blogs, and town hall meetings. Opponents of health care legislation said it revealed the real goals of the Democratic proposals. Advocates for health reform said it showed the depths to which their opponents would sink.[19]

The Internet offers up both a staggering amount of information and an equally staggering amount of untruth, from cyber-conspiracies to too-easy-to-be true ways to earn money by forwarding an e-mail message to friends. Slowly the Internet is supplying sites that strive to be reliable fact checkers and begin to serve somewhat as gatekeepers. One such site is Snopes.com, which is a popular fact-checking website run by David and Barbara Mikkelson. This site provides 48 categories ranging from autos to hurricane Katrina. Snopes.com tries to set the record straight on trivia items as well as serious political accusations. Is it a true news story that the average cost of rehabilitating a seal after the Exxon Valdez oil spill was $80,000 (not true), that Coca-Cola is an effective contraceptive (not true), that a woman died after visiting too many tanning salons in one day? Examples of misstatements about George Bush and Barack Obama illustrate political advertisements and attacks. The site affirms sadly what cultural critics have bemoaned for years: the rejection of nuance and facts that run contrary to one's point of view. "When you are looking for truth versus gossip, truth doesn't stand a chance," Barbara Mikkelson lamented.[20]

Another site is FactCheck.org, a politically oriented fact-checking site. Twenty-four-hour cable talking heads, partisan talk radio hosts, chain e-mails, blogs, and websites such as Worldnet-Daily or Daily Kos often transmit an undigested jumble of information and misinformation. Citizens need honest referees and fact-gatekeepers who can help sort out fact from fiction, especially in politics. FactCheck.org offers itself as a nonpartisan, nonprofit consumer advocate for voters. The site, a project of the Annenberg Public Policy Center of the University of Pennsylvania, seeks to reduce the level of deception and confusion in U.S. politics and "monitor the factual accuracy of what is said by major U.S. political players in the form of TV ads, debates, speeches, interviews and news releases."[21]

Even Wikipedia has had to apply some authoritative controls and credentialed experts because of inaccuracies and problems. Wikipedia has tens of thousands of amateur contributors of useful and easily accessible but sometimes unreliable content. Wikipedia started as a purely democratic, open online encyclopedia to which anyone could add content and comment without approval of a central authority. The logic was that the voice of a high school amateur had equal value to that of an Ivy League scholar or trained professional. As useful and as easily accessible as Wikipedia is, it represents the further replacement of knowledge by facts, and a distrust of informed expertise. Nevertheless, Wikipedia is one of the most visited sites (ranking about 17th), far outstripping Britannica.com, which has trained editors and careful review processes, but is ranked down at 5,128th on the list of most trafficked sites.[22] (Britannica.com, however, does cost $69.95 per month).

A blurring of reality and fantasy can be a positive condition, if it occurs in entertainment and play rather than politics. Reality television shows, which have taken over a swath of television, however, can be questioned as to artistic imagination, originality, and authenticity. Some reality TV players, for example in *The Bachelor* and also *The Bachelorette,* supposedly work as waitresses and salesmen and pharmaceutical reps but are really showbiz aspirants who pop up in other reality shows such as *Dancing with the Stars.*[23]

An area where reality appeared once to be solid was photography. But in the digital age, PhotoShop and other software can make fantasy seem real. This is true in film and computer games as well as traditional photography. Clever ads can show the impossible happening in them—talking cats, people morphing, infants behaving as adults, and so forth. Indeed, we are so accustomed to digital special effects, that we scarcely notice any but the more spectacular. Even photojournalism standards are changing, with more amateur photographers providing pictures that skew, interpret, or insert bias into a scene. Thus a photographer can go to a rally or demonstration and shoot photos that make it look as though 10 people showed up or 1,000—depending on the photographer's point of view.[24]

Fantasy sports, as a form of recreation, has enjoyed a noteworthy rise in popularity. Numerous sites online as well as a section of the magazine *Sports Illustrated* offer recommendations and advice for fans involved in fantasy leagues, baseball and football, in which fans use the names and statistics of real players but in an imagined league. The National Football League even has an official site where it provides tips, statistics, and information required for involvement.[25] A Google search of fantasy sites, however, yielded only half the number of Internet sites for baseball as it did for football, an interesting means for determining the reality of which fantasy is more popular.

Fourth Major Trend: Many Opinions, but Not without a Coarsening of Public Discourse

In spite of the blurring of news and entertainment, of a primacy of interpretation over fact-based evidence, there is a public conversation—or, at least, a lot of people are placing their opinions online.

As mentioned at the start of this chapter, there are more than 140 million blogs. BlogPulse.com claims there were more than 64,000 new blogs in the last 24 hours. BlogPulse.com and Technorati.com are key sites for navigating around blog sites of interest by category. Technorati.com has stopped counting but keeps track of about a million blogs, and provides a directory for blogs as well as a top 100 list. Technorati.com offers such categories as the total of all blogs (1,239,941), entertainment blogs (21,283, with subcategories of celebrity, film, music, television gaming, Anime, comics, books), next business blogs (20,047 with subcategories for finance, real estate, small business), and then sports blogs (8,917, again with different categories). After these, add politics, autos, technology, living (with subcategories of health, religion, food, arts, pets), green, and science.[26] (Technocrati.org, unrelentingly modern, has its conceptual organization rooted in the traditional library categories.)

Key blog sites provided different estimates of the total number of blogs, but no longer do bloggers seem to require any further confirmation of the health of the blogosphere by sheer numbers. Rather, energy and attention are on more specialized sites. Thus, for example, the Drudge-Report.com is a news aggregation website, generally conservative in tone. The HuffingtonPost.com site is a similar news aggregator, on the liberal/progressive end of the political spectrum. Perezhilton.com follows celebrities and gossip. Blog search engines are Technorati.com, BlogScope.com, and BlogPulse.com, which also list categories of blogs, provide current information, and track about 1,000,000 blogs. That adds up to a lot of people writing and offering opinions and observations.

In addition, it is estimated that 25 percent of Internet users each day spend time looking at YouTube. The estimates—no one knows the exact numbers—are that there are about 225 million video streams per day, with about 200,000 videos uploaded per day—which would take about 600 years to watch.[27]

Adding to this "flood of disconnected information . . . [this] curse of plentitude"[28] from the Internet are the images and data from television cable channels, as well as radio talk shows and social media. Such a flood of information can numb perception.

How to Get Attention

How do artists and marketers and entertainers break through this smog of huge amounts of data and infotainment?

Shock, pushing against societal limits, is one way of gaining attention. Artists and comedians will push and transgress social conventions. The borderline humor of 14-year old boys long ago crept into the mainstream via films, driven by commercial enterprise that finds a market for being lewd and risqué and outrageous, chipping away at institutional reluctance.[29] One performer, Ann Liv Young, uses dildos, urination, and masturbation as part of her theatrical act. At MoMa P.S., in February, 2010, and then in September, 2010, at the Issue Project Room, Brooklyn, she urinated and defecated in a one-woman show, "Cinderella."[30] Another comedienne has two million followers on Twitter, where, on any given day, one may find a picture she has posted of her dog defecating.[31] While the effort to be novel, original, and surprising as an attempt to grab attention is legitimate, often it merely is oafish and indecent.

Language that used to be regarded as obscene can be now found in such different areas as hip-hop lyrics, cultural bastions like the *New Yorker,* and blog sites. One prominent blogger claims to love those formerly repressed words that certainly do not adhere to traditional standards of journalism.[32] Talk radio practitioners like Howard Stern no longer shock with what used to be unacceptable language. "Obscene language is so prevalent and widespread as to be deprived of force . . . coarsening our concepts of what is and is not acceptable."[33] This is especially true in media that previously had some limits on coarse language.

Television advertisements are getting more frank and earthy. Stuart Elliott calls to attention ads for trucks, beer, and feminine hygiene products that use colloquial expressions for body parts, previously referred to by euphemisms.[34] The reason language in ads is becoming more coarse and vulgar, according to those who denounce the trend, is the niche-marketing nature of some cable channels. For example, the Oxygen channel aims its programming at younger women who presumably would not be as offended by the show's title ("Dance Your Ass Off") as older viewers. The same can be said for ads directed at younger men who are presumably less likely to object to sly double entendres about body parts. When advertisers have niche programming and marketing to work with, the burdens of making sure consumers are not offended is different. Consumers have to seek out niche programs and they know what they are getting; general programming has viewers who are not asking for or expecting trendier or objectionable advertisements or programming.[35]

The continued rise of individualism versus the decline of a sense of community may be good for freedom, but the common good suffers if no one is attending to community standards. But that is the issue: what are the norms and standards?

Fifth Major Trend: Huge Resources, but a Messy, Shifting, Disorganized Public Culture

Why is there this big, messy, chaotic, public cacophony? Are the new communication technologies really unleashing such a mixed stew of infotainment, political opinion, and scholarly resources? The range of available information is indeed very, very broad.

The nonhierarchical nature of the Internet plays a role in the dispersal of information. The U.S. National Science Foundation contributed to the formation of the Internet by overseeing the transition of the Internet from a government to a private, nonmilitary operation. The NSF "was forced to balance the competing visions of scientists, politicians, and private industry."[36] The result was the conflicts, trade-offs, and unexpected character of the Internet. Built into the very nature of the Internet is its democratic, chaotic character. Indeed mathematicians have attempted to calculate the degree of chaos—(the Lyapunov exponent)—in Internet traffic.[37]

Thus we have America, a large-scale, decentralized democracy, which has never been especially neat or orderly in approaching its problems, now engaged in a messy, complex process of finding out how to use these very powerful tools—without an instruction manual. It is stressful and messy, but also exciting.

Popular culture is the sum total of the images, ideas, sights, and sounds that inform us and entertain us daily. It is hard to keep tabs on popular culture anymore because of its kaleidoscopic, fragmented quality. The media spurt out streams of data about celebrities, new television programs, and new trends, as well as solid information. As channels of communication increase and multiply, there is an audience for each fragment. Smaller audiences seem more concentrated in their enthusiasm for narrower niches of the culture. Like-minded people tend to reinforce one another's opinions and may bar alternative views. Thus popular culture keeps fracturing into smaller and smaller slivers with an enthusiastic narrow band of consumers—rich in diversity in its own way. But as the cultural splintering increases, the communities of common interest grow smaller and more concentrated in fervor.[38]

Consumers then turn to sites that will aggregate and sort much of the information chaff for them and send alerts via e-mail or text messages. Cass Sunstein, a cultural commentator, points out our vast ability now to filter what we see, hear, and experience. The Internet and technological advances enable people to personalize their exposure to topics and points of view of their own choosing. We can allow in and we can filter out, with unprecedented powers of precision.[39] Sonicnet.com allows us to create our own musical world, named "Me Music." Zatso.net allows its members to decide what's news, select topics, and choose stories. TiVo, of course, gives us control over when we watch a favorite program. Other websites provide alerts and the means to lift chunks of content to our personal page.

As cultural unity becomes more fractured and diverse, it is harder to find themes and allusions that we share in common. The increasing cultural fragmentation of American society means the loss of a common vocabulary of allusion—a concern for writers and politicians who wonder if their audience will catch what they are trying to say.[40]

The fracturing of public culture has its price. Digital technologies put powerful communication tools into the hands of many, not all of whom practice conscientious journalism or have sufficient knowledge to be authoritative gatekeepers. Cultural critics of the Internet such as Andrew Keen lament the "flattening of culture" on the Internet. He raises an anguished cry about "endless digital . . . mediocrity . . . uninformed political commentary . . . unseemly home videos . . . [and] ubiquitous blogs have undermined our sense of what is true and what is false, what is real and what imaginary."[41]

Atomization of Music

Keen points to several areas where the digital realm exacts a price. The major change over the last 10 years has been how and where people listen to music. The shift to digital music seems to be the chief culprit of a massive change in how music is marketed and enjoyed. Keen laments, for example, the demise of Tower Records, an iconic chain of music retail stores. The bankruptcy and sale of Tower Records meant for Keen the loss of choice, knowledgeable salespeople, browsing, and serendipity of discovering a new album or new group.[42] While music piracy undoubtedly contributed to Tower Records' demise, the shift to online purchases of tracks or full CDs was the primary impetus. Perhaps there was no miracle strategy that could have kept Tower Records strong since the company was at the mercy of the record companies, who struggled to develop an online distribution strategy. Once the delivery infrastructure was in place that enabled people to sample and buy music online, it would only be a matter of time before retailers got whacked.[43]

The shift to digital has profoundly changed the music industry. At the end of last year, the music business was worth half of what it was 10 years ago, and the decline doesn't look like it will be slowing anytime soon. Total revenue from U.S. music sales and licensing plunged to $6.3 billion in 2009 from the 1999 revenue figure of $14.6 billion.[44] The 1990s witnessed an unnatural sales boost when listeners replaced their cassette tapes and vinyl records with CDs. In 1999, Napster, initially a free online file-sharing service, made its debut. Not only did Napster help change the way most people got music, it also lowered the price for a CD from $14 to free. Digital natives and digital immigrants can gain access to tens of thousands of songs for their MP3 players or iPods. Other areas of the music industry have also changed, such as a precipitous decline in ads on radio.[45] As David Goldman, the former head of Yahoo music, said, "The CD is still disappearing, and nothing is replacing it in entirety as a revenue generator."[46]

Music industry insiders see the music business as fragmented, lawless, and less and less profitable, yet flourishing. Although digital systems and digital distribution have replaced the old system of music distribution, and U.S. album sales and profits have declined in the last decade, there is more music out there. SoundScan tracked almost 100,000 new albums released in 2009, with millions of songs being downloaded for free on file-sharing systems like BitTorrent or swapped on social-networking sites like MySpace.[47]

Traditional symphony orchestras face these changes and are embracing "a host of new activities that would probably have left Toscanini with his mouth agape in amazement. Video projections . . . streaming performances live on the Internet or in movie theaters; allowing audiences to vote for encores through text messages; or establishing preconcert happy hours."[48] Orchestra executives are rethinking the role of the symphony in American society and trying to grapple with such questions as how to engage audiences reared in a digital, electronic, and social-networking world. As the president of the Los Angeles Philharmonic said, "It's a completely changing universe."[49]

Certainly there remain creativity and energy and new approaches among musicians and artists. I have spoken with small groups of performers and composers in the Boston area, who come together seeking new paths, trying innovative approaches.[50] They still keep their day jobs, however.

This is an another example of *creative destruction,* that painful process of competition and innovation in human activities. As I mentioned in chapter 4, the term describes the kind of disruptive revolution that occurs as a result of something new, which gives a decisive advantage to innovators and which strikes at the very existence of organizations that are failing to keep up with modern trends.[51]

Paradoxically, the fastest way to get information on changes in the music industry is via the Internet, the very cause of so much change. Indeed, that Tower Records went belly up so fast shows how rapidly industries can move online. Other brick-and-mortar retailers face similar challenges from nimble digital rivals. The video rental giant Blockbuster, with the growth of broadband and video-on-demand services, declared bankruptcy and planned to close "a sizable number" of stores.[52] A nimble upstart, Netflix, seems to have pushed brick-and-mortar Blockbuster toward extinction by using a website and the old-fashioned postal service to deliver movies to people's homes. Newspapers and newsmagazines struggle with the issue of how to charge people for reading their product online as traditional advertisers abandon papers for new outlets and innovative methods for reaching consumers. That might mean personalized ads sent directly to a consumer's smart phone or contextual advertising, the so-called one-to-one advertising, which aims ads at individuals on the basis of demographic information provided by consumers themselves unwittingly when they register at a popular website or for a special commercial offer.

Sixth Major Trend: Ever More Anxiety and Insecurity, Ever More Protections

No Longer a Glowing Future

Modernism once offered hope in the future because technology would bring a better life. Indeed, in 1939 the New York World's Fair promised "a future in which life in this country for the great majority would be vastly richer and easier."[53] Now, for multiple reasons, especially economic,[54] people worry about where all of this will end. Media information about the world somehow being unsafe fosters an increase in the need for passwords, picture identification, and card-access-only locked rooms.

The world does contain dangers, but modern technologies can amplify our sense of vulnerability. There are bad guys out there who kill people and have a terrorist or criminal intention. But the very increase in security measures seems to mark a greater rise in our sense of exposure. Airport security and metal detectors in courthouses and government buildings are the inconveniences of necessary security. Especially annoying are the defenses required against Internet viruses and malware that can attack our computers. New companies and websites, such as Norton and McAfee, rush to offer protection, including warnings to parents to protect their children from online dangers.

Social-networking users are now more alert to privacy breaches. Privacy is a concern when one spends time on social-networking sites. Facebook and MySpace have gotten into hot water by selling data to advertising companies that could use consumers' names and personal details, even though the sites promise not to share such information without consent.[55] Also harvested online are *cookies,* those packets of digital data that our computers pick up and which can be used to trace our online purchases and browsing preferences.

On a more positive note, modern communications can both inform about the spread of disease and help coordinate a global response. Such was the case with SARS (severe acute respiratory syndrome), which spread rapidly around the world via people traveling by jet. Within a matter of weeks SARS spread from the Guangdong province of China to rapidly infect individuals in some 37 countries around the world. Although it killed around 1,000 individuals, it did not lead to the devastating health impact that many feared.[56]

Similar concerns are raised about invasive species of plants and insects, which can spread wider and wider, causing native organisms to go extinct or alter their traditional habitats. The science writer Richard Preston writes of invasive Asian longhorn beetles, the Ebola virus, and deadly fungi. To describe the collapse of traditional defenses against such environmental threats he offers the image of a sort of palace for the earth's biosphere. Each palace room has its own décor and unique inhabitants; many of the rooms have been sealed off for millions of years. "Now the doors in the palace have been flung open, and the [defensive] walls are coming down."[57]

A way to quantify increased concerns about security is to look at the growth in the use of security cameras. In part, improved technologies of video cameras render them more reliable and capable of functioning in low light. In New York City's financial district in 1998, when the New York City Civil Liberties Union started a count, there were 769 cameras; in 2005 the figure was 4,468.[58]

The rise in the number and popularity of gated communities, mentioned above in the section on divided values, also indicates vulnerability and anxiety. Writing two years before 9/11, Blakely and Snyder in *Fortress America* describe a growing fortress mentality among Americans who seek "gates, fences, and private security guards" in the face of rapid change.[59] "There is a growing fear about the future of America. Many feel vulnerable, unsure of their place."[60]

One wonders if a new symbol of America is the gated community. The popularity of gated communities has increased nationwide, especially since 9/11. In some states up to 40 percent of new homes and about 6 percent of the national total of households are behind walls in the quest for safety and security.[61] As Edward Blakely and Mary Gail Snyder say in *Fortress America*, "The setting of boundaries is always a political act. Boundaries determine membership; someone must be inside and someone outside."[62]

For comparison's sake, we can contrast the current somber view with a far more optimistic view in an earlier, different time. At the end of the 1930s America was still grappling with the residue of the Great Depression and Hitler had invaded Poland to start a world war. David Gelernter, a historian of the 1939 New York World's Fair, describes the optimism in the air at that time. Using Gallup and Roper polls from 1939 and 1940, Gelernter found that a plurality of Americans were optimistic about the future. Gelernter finds their optimism surprising, since "we hold a much bleaker view today . . . when . . . our world is incomparably better off than 1939's."[63]

Conclusion

America is undergoing an enormous, demanding experiment. We are undergoing a test of our adaptability to new technologies and a test of whether or not we can gather the political will to make the hard choices to deal with unresolved issues of cultural values, the environment, and the economy. Which technological advances, innovations, and social changes work and which do not? No authority is in charge of this testing; the process is messy and evokes anger and anxiety.

The United States is not falling apart, but it is having a hard time. Politics is a no-holds-barred brawl, and the national economy does not grapple very well with global competitors. Change and conflicting values increase stress and intensify the search for remedies. When accustomed patterns break down, people feel disoriented, deprived, and anxious. Social transition and social dissatisfaction can lead to extremism and cultural wars.

Whose culture is it? People join movements if they are fearful of deterioration in the quality of their lives with the likelihood of further declines. Social friction increases as one subculture reacts to different systems of values. A changing social order creates disorientation and anxiety.[64]

America's love of technology and innovation and its lack of the long traditions of Europe not only foster but even encourage grand and risky experiment. Andrew Keen suggests a metaphor for the role of change (at least in the music industry): the forest fire that clears all for vigorous new growth.[65] But that seems too destructive. A better metaphor might be the automotive industry. The history of the auto industry in the United

States was initially one of broad innovation and experimentation as well as competition. There were more than 1,800 manufacturers during the period from 1900 to 1930.[66] Through market and economic forces, the number was winnowed down and consolidated, with the emergence of the Big Three (Ford, General Motors, and Chrysler, with several smaller firms) by the middle of the last century.

Of course, the Big Three have also had their trials with global competition, and thus demonstrate the limits—and dark side—of this comparison. But if the government intervention in the auto industry serves the larger good, the lesson of this example then is that we are not powerless in the face of change but can nudge the forces of change into positive directions.

There is a constant, uncertain process of evolution. The best may not always survive and some good things will be lost. Change is relentless, and things will be different.

9

Conclusion

What is the significance of the changes looked at in the previous chapters? What kind of toolbox of skills does a person need to successfully meet the contemporary world of rapid change? Are there theoretical ways of thinking about change? What gets lost amidst change?

The Revolution Is Here and Not Going Away

The big engines of contemporary change are not going away. The computer and the Internet, the digital and information revolutions, offer too many benefits at every level of society. These technologies spark new industries and create new efficiencies.

Also, smart phones—really, handheld computers—iPads, and computer tablets are just downright fun to use. Such devices with fabulous screens enable us to connect and explore in new ways. For example, art lovers can browse museums around the world or search for reproductions on the Web and have them at one's desk instantly. Music lovers can find and download thousands of tracks. Researchers can reach into the archives of great libraries or search electronic journals. Do you want to know the latest stock market figures? Do you need a GPS to find that new restaurant? Need to access your e-mail? Too many advantages and pleasures flow from the new digital and communication technologies. They are not going away. There is delight in something that works well and does more than one could ever have dreamed of. Testimonials abound as to their uncanny pleasures. "I had seen the future—and it was beautiful," says Ann Kirschner.[1] Virginia Heffernan enthuses about her iPad: "It's uncanny. Once yielded to, it's scintillating for being uncanny."[2] Early desktop computers initially were regarded as toys and used for games or expensive typewriters until people began to realize their potential. Then rapidly computers began to perform better those tasks that used to be more tedious and less efficient—crunching large amounts of data and linking to share information.

The dark side is not these powerful tools themselves but rather the disruption and uncertainty they bring in their wake as society and people are changed by the abundance and speed of information. Some people will be left behind in a more competitive world. Some good features of the past may be lost—only temporarily, it is hoped—until we tame these powerful resources. Uninterrupted leisure, the ability to escape, sustained thinking, privacy, all need to be more consciously valued and practiced more deliberately. A new age of anxiety requires filters to manage and make sense of the blizzard of information that comes our way.

Change brings trade-offs. Change introduces speed as both a convenience and a stressor. New technologies make multitasking both possible and a potential source of inefficiency. An abundance of information both enriches and overwhelms.

How Do Successful People Manage Change?

Look Change in the Eye

I do not particularly like change.

A mental approach, however, that helps me manage change is to be intentionally aware of change and look it in the eye for what it is. A changing environment requires us to have a focusing lens, a consciousness of the nature of the flux in which we find ourselves. This heightened awareness enables us to anticipate change, prepare for it, make choices about it, direct it, learn from it, and—possibly—maybe even enjoy it. Sometimes approaching chaotic change as a game prompts our creativity.[3]

Change is ubiquitous and provides opportunities for both disruption and growth. Life is not always the way we'd like it to be or the way it is supposed to be. Change is a constant and some individuals and some organizations deal with change better than others. The organizations that seem to succeed tend to be "self-aware, that is to say they are conscious, are better able to see change coming, more honest in determining how best to respond to it, and more courageous in taking action steps to capitalize on change itself."[4] They seem to do this best by taking care of human factors rather than merely technological factors, helping the people within the organization to be conscious of change and its dual potential—for opportunity or for disruption.

Managers and leaders must think purposefully about change. Organizations are increasingly expected to find better and faster ways of responding to the challenges brought by the revolution in information technology, globalization, smarter customers, and a changing employee base. "[T]he world is in a constant state of change, and no organization, in the United States or elsewhere, can escape the effects of operating in a continually dynamic evolving landscape. The forces of change are so great that the future success, indeed the very survival, of thousands of organizations depends on how well they respond to change or, optimally, where they can actually stay ahead of change."[5]

History is littered with those who failed to adapt to change. Those old-time door-to-door salesman were successful as long as women were at home. When women went out into the work force, retailers had to change—stores extended hours to catch customers who worked during the normal work hours. Iconic glossy magazines like *Life* and *Saturday Evening Post* did well until advertisers realized television offered a bigger audience for a lower cost. Publications that did not adapt did not survive.[6] Newspapers are scrambling to find ways to get consumers of news to pay for online information. The convenience of shopping on the Internet unleashes an unimaginable abundance of online retailing. The challenge for all, retailer and consumer, is to adjust to the changes, even if we feel we don't have much control or don't particularly like what's happening.

Seeing change as change—rather than as a threat—is to be forearmed. Reframing the challenges will foster a creative response.

The Goal Must Be to Improve the Quality of Life

Many people that I interviewed in the course of preparing this book mentioned how overwhelmed they frequently were, how they can't seem to catch up, how busy they were.

Self-help books offer useful tips about managing time and clarifying our goals. For me, a core task is to confront the question of what is the good life. Can we remind ourselves to make room for breathing, for the people and the activities that mean the most to us? We do not want to lose our sense of self—so easy to do in the race to get it all done. Excessive haste can lead to errors and loss of valuable time.

Valorie Burton, an adviser on lifestyles, reminds her readers to live in the present and try to be engaged as fully as possible in what it is we are doing. When we are "off from work," we should give our time solely to nonwork-related activities—to eat regularly, preferably sitting down, to exercise regularly, preferably standing up, to take our rest seriously, and to remember to have fun, at least once a week.[7]

What is the balanced life? The relentless time pressures of modernity require us to push back against some of the demands. Being alert to our own pattern of responses can nurture interiority rather than outer-directedness. We must realize that we have choices how we spend our time and how we think about our involvement in work. We have the power to make choices that favor family and friends and still be committed to a career. There is enough time to do the really important things.[8] The task is to clarify what is important to us, then prioritize, and then make the necessary compromises.

Sabbatical from E-mail and Texting

Information is good and now so abundant. The challenge is to manage information, make it useful for us and not be overwhelmed by the wealth of data that is available to us. Several vivid phrases express the value of taking short sabbaticals from technology or, so to speak, going off the grid for a time without e-mail or Web access. A *Walden zone*[9] suggests cutting down on stimulation, whether from the phone or radio or iPod. Time off the electronic grid increases our awareness of our lives, similar to the refreshed perception we have of home on our return from traveling.

We need to find a way to filter the flow of stimulation from modern communication devices. Do we trust the new gatekeepers—whether they are websites that aggregate news and send us alerts or traditional news gathering organizations such as the *New York Times* and National Public Radio?

Increasingly people are reporting on their brief experiences of going off the electronic grid. Freelance writer C. M. Boots-Faubert agreed to an experiment. He could use technology existing prior to 1980, but not later, so therefore: no computer, no cell phone, no laptop. To write he had to resurrect an old IBM Selectric typewriter, which he soon realized did not make hyperlink connections, nor did it have a find-and-replace function key or a spell checker, nor could he easily change fonts to bold or italics or underscore. He was burdened with finding a pay phone and he had no automatic reminders popping up in the task bar of his notebook computer screen.[10]

Boots-Faubert was startled to realize he enjoys some quiet pleasures other than the computer game that sits unused on his desk. He reports that he learned something important: how hard it is to say no to technology. Because "modern technology is tightly woven into the very fabric of our society," he says, "I can survive without modern technology, I just can not live without it. No longer do we dial into the Internet when we needed it, but it is now the always-on constant source of information."[11]

College students report similar angst when for a class assignment they have to do without their phones and texting for a day. Some colleagues have told me they are assigning students projects that involve turning off their cell phones for a day or even—horrors—a week. Students have to confront themselves but also learn that perhaps they are addicted to the stimulation of their phones and texting. A study from the University of Maryland reports college students who swore off social media showed signs of withdrawal similar to those of drug addicts going cold turkey.[12]

Care of the Self

Every culture contains within itself some resources for dealing with challenges within that culture. People, online sites, books, all offer advice. The more I looked, the more resources I found—whether it was for a personal trainer, yoga classes, stress reduction, or support groups of many kinds.

The simplest resources are the most effective. I found that when I went for a jog, unplugged and able to leave behind all electronic devices, I had a quiet mind. Clearing my mind and emptying it seemed to allow solutions to bubble up to the surface and an open moment to appreciate many of the good features in my life. Setting firmer boundaries between work and nonwork, between work and vacation, refreshes the spirit.

Unemployment, worries about finances, limitations of health, all diminish the zest for life. One has to get through these hard periods however one is able.

William Powers[13] believes that while the digital age is still young, we should be thinking seriously how we will be managing our technologies of connection. He sings the praises of paper and books as the old tools of the technology of inwardness, of zones of clarity and calm. Powers reminds us that we can indeed regulate the quality of our experience, especially by striking a healthy balance between connected and disconnected, the outward life and inward one.

Keeping an Open Mind

In a world of many specialties and niches, it is useful to keep eye and ear open to what is unfamiliar or challenges our own thinking. If we come to terms with the fact that modern reality is complex and messy, we can be receptive to a diversity of thought. Institutions perform better, as author Carl R. Sunstein puts it, "when challenges are frequent, when people do not stifle themselves, and when information flows freely."[14] We all need accurate information to stay on course and correct errors.

Institutions will be more productive and successful if they subject leaders to critical scrutiny and if courses of action face monitoring and review. Diversity of thought and dissent can reduce the risks of effort that come from group polarization. Diversity and dissent can actually serve a useful function of making sure ideas are thoroughly explored. Well-functioning individuals and groups try to entertain a diversity of views, if only to protect themselves against blunders and confusion. Sunstein also comments, "If teams of doctors want to make accurate diagnoses, they will promote a norm of skepticism."[15] The problem of division arises from the increase in extremity of views when like-minded people talk to one another; this is true in politics, religion, investing, and corporate leadership. Cultists and extremists tend to be separated from the rest of society.

Adaptation, Adaptation, Adaptation

Successful people—and nations—surmount challenges. And the challenges brought by change require adaptation.

In an interview with a customer service representative on a Medicare website I learned that more and more retirees wanted, and were using, Internet access. This was not true a few years ago with an older cohort. A newer generation of seniors, digital immigrants, having enough experience with the digital revolution, now want continued electronic access to institutions that have been serving them but which until recently did not offer interactive websites.

The power of globalization compels change. An early example is foot binding in China. This practice persisted for a thousand years but was eradicated within a generation from 1910 through the 1920s because of exposure to the values of the rest of the world.

Globalization can produce homogeneity but also can be a threat to local homogeneity. In parts of the Third World local people can talk about Ronaldo, Mike Tyson, and Madonna as they drink a Guinness or Coca-Cola. Some enclaves of homogeneity, such as the Amish in Pennsylvania, are less distinctive than they once were, but they can still feel the loss of difference and may feel that their identities are threatened. Why? Because their world is changing and some of them don't like it.[16]

Still Evolving, but How Fast Can We Change?

Change produces change. We are being changed. The younger generation is different from the earlier generation because they are stepping into the flowing river further downstream and did not experience what caused the turbulence upstream.

We are still evolving, but not as fast as expected.

Our species spread out of Africa about 60,000 years ago and encountered environmental challenges that they could not overcome with their level of technology. There were challenges of disease, high altitudes, cold climates, wild animals. Presumably there were genetic mutations over time that spread by natural section, that is, those who carried the mutations would have more children than those who didn't. While we have a few examples of very rapid natural selection, most natural selection seems to have occurred at a far slower pace than envisioned by researchers.[17]

An example of a rapid genetic change as a response to the environment is the capacity of some people, specifically Tibetans, to live in the harsh conditions of low oxygen at high altitude. This gene variant, which adjusts red blood cell production to carry more blood oxygen, appears to have originated and spread within the last 3,000 years. This biological adaptation is in addition to technological adaptations to high altitude, such as suitable clothing for cold weather. Two other of the very few examples of strong, rapid natural selection are lighter skin tone adapting to reduced sunlight in northern climates and the adaptation for the lactase enzyme that digests lactose, enabling people to digest milk. These two genetic changes appear, however, to have occurred over a period of 5,000–10,000 years.[18] The pressures humans face now are of our own making—an altered environment, population crowding, and the complexities brought about by our own technological innovations—and we don't have a few thousand years to adapt to these. Author Mark C. Taylor expresses it this way: "We are currently living in a moment of extraordinary complexity when systems and structures that have long organized life are changing at an unprecedented rate. Such rapid and pervasive change creates the need to develop new ways of understanding the world and of interpreting our experience."[19]

Searching for New Paradigms

There are many different kinds of change, including changes that affect individuals and changes that affect large groups. Technology, economics, and politics all pull personal and national changes in their wake. A key property of all change, however, is linkage. No change exists in isolation, but has systemic connections, so that when one thing is altered other events follow.

Mark C. Taylor, seeking theoretical ways to understand the changes we are living through, speaks of blurring categories and the collapse of metaphoric walls. He says: "Apparently unrelated developments, which had been gradually unfolding for years, suddenly converge to create changes that are as disruptive as they are creative. We are currently living in a moment of extraordinary complexity when [economic, political, social] systems and structures that have long organized life are changing at an unprecedented rate. Such rapid and pervasive change creates the need to develop new ways of understanding the world and of interpreting our experience."[20]

Taylor differentiates complexity from catastrophes and chaos. Catastrophes are singular events producing changes that issue in sudden dislocations and disruptions. Chaos creates situations that are disproportionate to their causes in such a way as to make it impossible to predict and prepare for similar future events. Complexity theory seeks to track the movement from lesser to greater complexity, a discontinuous change between too much and too little order.[21]

Other studies take a more catastrophic approach to national change, specifically when complex societies collapse. Joseph Tainter, studying historic patterns of collapse, sees economic causes as the central explanation for serious national change and decline, even though external shocks such as invasions, crop failures, or disease may be the apparent causes. Thriving societies invest in energy, education, and technological innovation. The globalized modern world is subject to many of the same stresses that brought older societies to ruin.[22] Jared Diamond suggests that some societies failed to adapt their practices to warning signs of environmental degradation and collapse.[23]

Thomas Friedman, a columnist and economist concerned about America's economic decisions, writes: "Nations thrive or languish usually not because of one big bad decision, but because of thousands of small bad ones—decisions where priorities get lost and resources misallocated so that the nation's full potential can't be nurtured and it ends up being less than the sum of its parts."[24]

The challenge is to find a balance between change and stability. If there were no resistance to change, society would have trouble holding itself together, says Douglas Rushkoff. With no identifiable constants, our world would feel too fluid, too irregular, and too chaotic for any meaningful interactions. Our era is indeed a period of extreme novelty.[25] Novelty can prepare the way for creativity and new directions.

People ordinarily accept change only out of necessity. We come to terms with the inevitable when the way we've been understanding the world so contradicts our felt experience that it stops serving any useful purpose. A model or paradigm of reality can be thought of as working when it has the ability to explain the present or predict the future with some degree of success. It reassures us that everything really is all right and proceeding on schedule.

The risk is mental rigidity. As individuals and a society we must be cognitively flexible during times of change. Discomfort with complexity and confusion when rules are not clear can lead to dangerous mental errors. It only compounds errors when we seek simplistic solutions to complex problems and demonize those who disagree with us.

Schemas are cognitive ways of understanding the world and interpreting our experiences. Thus some people might attempt to explain everything as caused by fate or God's will or political enemies. But overly childish or monodimensional thinking does not do justice to complex realities.

Periodically, simplistic political groups or religious fundamentalists seek refuge from, or solutions to, the complexity and demands of the era. Such groups may seek comfort and reassurance in authoritative scripture and or an imagined golden past. Too often such groups and individuals interpret events in ways that confirm their preexisting beliefs while ignoring ambiguities and facts.

Integrating Technologies

Technologies are tools that extend or enlarge human capacities. Not all technologies are good. But often we can't tell that, except in hindsight. Thus, when the British Royal Navy launched the *Dreadnought* in 1905, this battleship was judged to be technologically supreme, faster and more heavily armed than any ship afloat, outrunning and outgunning any possible opponent. The problem was that this weapon made Germany so anxious that Germany made the decision to build its own dreadnoughts. An expensive arms race began, culminating in World War I. Britain's naval supremacy and technological lead vanished, without the *Dreadnought* ever playing a significant role in the war.[26]

The lesson? Probably we need a variety and range of innovations—and the solid judgment to assess the costs and rewards of each. Often market forces serve this function. Some innovations are remarkably useful and keep getting improved over time until they are outmoded. Thus the computer has replaced the typewriter. The internal combustion engine, so useful in cars for more than a century, is gradually and ever so slowly being replaced by electric hybrids. Desktop computers also began as relatively simple machines, kits off the shelf for the technically savvy. Cars and computers have evolved and diversified toward greater complexity. When they work as designed they are marvels of elegance and usefulness. However, when glitches show up in the complex structures that we humans create, they frustrate and torment us.

Technological lock-in is an idea used in economics to describe cases where an established but second-best technology continues to dominate because of its advantages. Thus an older technology, such as the QWERTY keyboard, seatbelts, or the VHS over the BETA video recording may have continued because of costs of retraining, or costs of new systems, or sheer convenience.[27]

Even good ideas are not always quickly adapted. Bernard D. Sadow invented luggage on wheels in 1970, getting the idea after he lugged two heavy bags through an airport. His idea, four wheels on the bottom of a large bag that could be pulled, however, did not take off immediately. The 1970s were a time when airplanes were decisively replacing trains for long-distance travel, and people were dealing with luggage without porters or bellhops. A better wheeled piece of luggage, Rollarboard, was invented in 1987 by Robert Plath. Rollarboard added two wheels and a telescoping handle to suitcases that rolled upright, rather than being towed flat like Mr. Sadow's four-wheeled models. In this case the early adopters were flight attendants.[28]

Another innovation that people initially reviled and regarded as a threat was in vitro fertilization. Now about 25 years old, IVF has been so successful it is almost routine. Gradually the public came to accept this medical advance after it was proven safe and not some scientific nightmare.[29]

Change Involves Losses, Even of Valuable Qualities

Change does involve loss. Nostalgia tends to emphasize what was good and push to the side what was inferior. This book is not about nostalgia, but nostalgia does have its appeal.

The world, and human machines, were once simpler. Memory distorts and tends to recall the rosy and forget the dark. Looking back at the past will bring out the complexity, the "both and" quality of every age. Each era faces its difficulties, and each decade also offers special pleasures to revisit and examples of valuable qualities we do not want to lose.

Pete Hamel describes the sense of community and connectedness where he grew up in Brooklyn during the 1940s and 1950s. On summer evenings middle-class neighbors would sit on stoops and watch boys play street ball and girls roller skate or skip rope. "There was no television then, so they made their own entertainments."[30] Things changed and the streets grew empty at night when television pulled people inside, perhaps distracting them from their boredom and the shadow of possible nuclear war, which Hamel passes over.

Bill Bryson's memoir of the 1950s is upbeat and positive. "I can't imagine there has ever been a more gratifying time or place to be alive than America in the 1950s. No country had ever known such prosperity."[31] Bryson sketches the wealth of the country then: the United States owned 80 percent of the world's electrical goods, two-thirds of the world's productive capacity, more wealth than the other 95 percent of the rest of the world combined, and 60 percent of the oil, in addition to which 99 percent of the cars in America were made in America. "No wonder people were happy. Suddenly [after the Second World War] people were able to have things they had never dreamed of having."[32] Bryson goes on, "Baseball in those days dominated the American psyche in a way that can scarcely be imagined now,"[33] and elsewhere adds, "Men wore hats and ties almost everywhere. Women prepared every meal more or less from scratch. Milk came in bottles."[34]

The pace of life was slower and one could find solitude and time for reflection. Historian Liaquat Ahamed describes the struggle to find solutions to the economic crisis of the 1920s. The powerful bankers at the time, relying on the relatively primitive tools and sources of information of that era, communicated by mail—at a time when a letter from New York to London took a week to arrive—or, in the situation of real urgency, cable. The way to travel was not by plane but ocean liner and it took five days to cross the Atlantic. This allowed for reflection before a response was sent. "It was an era when Benjamin Strong, head of the New York Federal

Reserve, could disappear to Europe for four months without raising too many eyebrows."[35]

Perhaps the most anguished description of radical cultural change comes from Austrian author Stefan Zweig. His life bridged the secure, peaceful world of pre-World War I and the horrors of World War II. In his autobiography he expresses his bewilderment at how different his life is from the lives of his father and grandfathers, "entirely separate worlds."[36] He marvels at how much has been compressed into life during the 20th century, while his father and grandfather lived lives that scarcely had any disturbance, anxiety or "noticeable transitions" (p. vii).

Zweig grew up in Austria, prior to World War I, when life in Austria, under a millennium-long monarchy, rested firmly on permanency and stolid tradition. Everything had its time and place, and one could look forward to a particular day in the future when one could retire at a specific pension. "It was an ordered world with definite classes and calm transitions, a world without haste" (25). The bourgeois lived lives of comfort, pleasure, stability, and security and "nothing unexpected ever occurred" (26). Speed was regarded as both unrefined and unnecessary.

Zweig recalls the dullness and dreariness of his youthful world. The authority of teachers and parents was unquestioned. Students had to respect "the existing as perfect, the opinion of the teacher as infallible" (34). Elders believed that youth "were not to have things too easy" (34). It was not a good time for females. Families kept rigid control of girls, how they dressed and were educated. The mania for sports had not yet reached Europe from England to distract and engage the young. Sex was not mentioned. Young people being together without the strictest supervision was "unthinkable" (74).

A cry of the heart, an acute sense of loss, is expressed by author Alvin Kernan,[37] who laments the decline of "high-culture assumptions about canonical perfect works, imaginative geniuses, single, fixed meanings in texts" (290) and the sense of continuity with earlier generations in the "shared, continuing search for truth" (290). "The ambiguities, irony and complex structures of thought fostered by print now seem increasingly superfluous. Where the fixity of the printed book encouraged the conception of masterworks and permanent human truths, databases and hypertexts and TV programs that flicker past make originality, form and permanence seem very quaint" (244). He ruefully acknowledges, "There is no reason to think that the electronic era will not be similarly responsive to human needs, but what it will produce will be different. It is not that bad is replacing good but rather something new is replacing the old" (243).

Other writers[38] wonder at the seeming loss of beauty in modern music and art. Certainly in both of these arts the challenges are different, as well the economics and the experience. Artists do not live isolated from technology and technology's unforeseen consequences. We rarely listen to the same songs, and we rarely sing spontaneously or together except, perhaps, in church. If kids do any singing, at camp, for instance, the only songs they know in common are television jingles.

The Author's Paradigm

Change is opportunity. American culture, at its best, has been optimistic and hopeful. If change and progress bring a better, fuller life, it is to be embraced. America has usually tolerated diversity and a broad range of options in order to discover what succeeds. Some things should change: those that diminish the quality of life, lessen freedom, cause loss of privacy, and shrink us and demean the spirit.

Life without change is stifling and boring. The challenge is to enable our best selves to come forth.

A positive mental stance toward the new makes life interesting. After all, we seek novelty when we watch a new movie rather than repetitiously watching the same one.

Evolution confirms that successful organisms can adapt to moderate change. Too much change and there is a tipping point into discontinuity, a break; too little change leads to dullness.

The good old days were not always so good. Most people, the nonrich, faced a life of toil and more toil, of being cold in winter and hot in summer. Food was not always varied nor abundant. People died young, and often faced boredom and exhaustion.

In spite of losses and uncertainty during this decade, there are many reasons to be pleased about living in the present era—with so many more opportunities, the possibility of a longer life, an enrichment of experience via extraordinary media, and the efficient means for staying in contact with family and friends. We are alive during an extraordinary time.

Notes

Chapter 1

1. Robert Darnton, "The Library in the New Age," *New York Review of Books,* June 12, 2008, 72–74.
2. Darnton, 72.
3. Gordon E. Moore, "Cramming More Components onto Integrated Circuits," *Electronics* 38 (April 19, 1965), http://download.intel.com/museum/Moores_Law/Articles-Press_Releases/Gordon_Moore_1965_Article.pdf (accessed February 1, 2010).
4. Martin LaMonica, "IBM Software Boosts Information on Demand," February 13, 2007, http://news.cnet.com/IBM-software-boosts-info-on-demand/2100–7345_3–6159025.html?tag=mncol (accessed May 12, 2009).
5. Kevin Kelly, "The Speed of Information," http://www.kk.org/thetechnium/archives/2006/02/the_speed_of_in.php (accessed February 7, 2009).
6. Thomas H. Davenport and John C. Beck, *The Attention Economy: Understanding the New Currency of Business* (Boston: Harvard Business School Press, 2001), 2–3; Torkel Klingberg, *The Overflowing Brain: Information Overload and the Limits of Working Memory* (New York: Oxford University Press, 2009), ix.
7. See Stephen Kern, *The Culture of Time and Space: 1880–1918* (Cambridge, Mass.: Harvard University Press, 1983), 118; John Jervis, *Exploring the Modern* (Malden, Mass.: Blackwell Publishers, 1998).
8. Bobbie7-ga, "American Advertisement in the Media," August 2002, http://answers.google.com/answers/threadview?id=56750 (accessed

April 9, 2010); Michael Brower and Warren Leon, "Practical Advice from the Union of Concerned Scientists," http://www.ucsusa.org/publications/guide.ch1.html (accessed April 12, 2010).

9. "Number of Jobs Held, Labor Market Activity, and Earnings Growth among the Youngest Baby Boomers: Results from a Longitudinal Survey," http://www.bls.gov/news.release/pdf/nlsoy.pdf June 27, 2008 (accessed June 1, 2010).
10. Alison Lobron, "Is Unlimited Vacation a Good Thing?" *Boston Globe Sunday Magazine,* July 20, 2008, 8.
11. Chris Anderson, *The Long Tail: Why the Future of Business Is Selling Less of More* (New York: Hyperion, 2006), 4.
12. Anderson, 7.
13. Rebekah Nathan, *My Freshman Year: What a Professor Learned by Becoming a Student* (New York: Penguin, 2005), 38, 44.
14. Robert D. Putnam, *Bowling Alone: The Collapse and Revival of American Community* (New York: Simon & Schuster, 2000).
15. Allen Guttman, *From Ritual to Record: The Nature of Modern Sports* (New York: Columbia University Press, 1998, 2004) and Martin E. Marty, "Ritual in Sports, Sports as Ritual," *Park Ridge Center Bulletin,* August 1998, 19, http://www.parkridgecenter.org/Page127.html (accessed October 26, 2010).
16. John P. Robinson and Geoffrey Godbey, *Time for Life: Surprising Ways Americans Use their Time* (University Park: Pennsylvania State University Press, 1997), 47.
17. Paul B. Brown, "Wireless Codependency," *New York Times,* February 17, 2007, B5. See also JohnatmyITforum.com, "Wrist Watch Sales Continue to Decline," August 27, 2006, http://myitforum.com/cs2/blogs/jgormly/archive/2006/08/27/Wrist-Watch-sales-continue-to-decline.aspx (accessed April 14, 2010).
18. *The Structure of Scientific Revolutions* (Chicago: University of Chicago Press, 1962).
19. "Pole Wakes Up from 19-year Coma in Democratic Country," *Time* magazine, June 18, 2007, 18.
20. "Pole Wakes Up," 18.

21. United States Postal Service, "First-Class Mail Volume Continues to Decline," Press release, www.usps.com/communications/news/press/2003/pr03_048.txt (accessed June 8, 2010).

Chapter 2

1. Enquiro Research, "The Rise of the Digital Native," http://www.enquiro.com/whitepapers/pdf/The-Rise-of-the-Digital-Native.pdf (accessed February 12, 2010).
2. Robert Darnton, "The Library in the New Age," *New York Review of Books,* June 12, 2008, 72–73.
3. Anne Helmond, "How Many Blogs Are There? Is Someone Still Counting?" February 11, 2008, http://www.blogherald.com/2008/02/11/how-many-blogs-are-there-is-someone-still-counting/ (accessed August, 2, 2010).
4. "46 Million Websites Added to Internet in January-April 2009 Study," May 11, 2009, http://www.alootechie.com/content/46-million-websites-added-internet-jan-apr-2009-study (accessed May 10, 2010); Pingdom, "Internet 2009 in Numbers, January 22nd, 2010, http://royal.pingdom.com/2010/01/22/internet-2009-in-numbers/ (accessed September 10, 2010).
5. Tom Boutell, "How Many Web Sites Are There?" http://www.boutell.com/newfaq/misc/sizeofweb.html; and The Age, http://businessnetwork.theage.com.au/articles/2005/11/18/3491.html (accessed June 12, 2008).
6. Wisegeek, "How Big is the Internet?" http://www.wisegeek.com/how-big-is-the-internet.htm (accessed March 7, 2010).
7. http://royal.pingdom.com/2010/01/22/internet-2009-in-numbers/ (accessed September 10, 2010); http://twitpic.com/t18ml (accessed March 7, 2010).
8. Say Keng Lee, "Human Knowledge Is Doubling Every 32 Minutes," http://optimumperformancetechnologies.blogspot.com/2007/11/human-knowledge-is-doubling-every-32.html; and http://leitl.org/sci.nano/5165.html (accessed July 27, 2008).
9. Richard Wray, "Internet Data Heads for 500bn Gigabytes," Guardian.Co.Uk. (Monday, May 18, 2009), http://www.guardian.co.uk/

business/2009/may/18/digital-content-expansion. (accessed March 12, 2010).

10. John Markoff, "Striving to Map the Shape-Shifting Net," *New York Times,* March 2, 2010, sec. D, 1 & 4.
11. Richard Saul Wurman, *Information Anxiety* (New York: Doubleday, 1989), 32.
12. Guo-Qing Zhang et al., "Evolution of the Internet and Its Cores," *New Journal of Physics* 10 (2008): 123–27 (Digital Object Identifier No. doi:10.1088/1367–2630/10/12/123027, published December 18, 2008; accessed March 2, 2010).
13. Todd Oppenheimer, "Reality Bytes: We Listen in on the New-Media Moguls—and They're Nervous," *Columbia Journalism Review* 35 (1996): 40–46; and http://www.lessonsforliving.com/speed.htm (accessed June 10, 2008).
14. Google Books Library Project, Google.com, http://books.google.com/googlebooks/partners.html (accessed February 2, 2009).
15. Darnton, 78–79; Conrad de Aenlle, "Digital Archivist, Now in Demand, " *New York Times Sunday Business,* February 8, 2009, 15.
16. Brian Stelter, "C-Span Puts Full Archives on the Web," *New York Times,* March 16, 2010, sec. C, 1 & 6.
17. Pew Research Center Publications, "Updated Demographics for Internet, Broadband and Wireless Users," http://pewresearch.org/pubs/1454/demographic-profiles-internet-broadband-cell-phone-wireless-users (accessed April 7, 2010).
18. *Current Population Survey*, posted Oct 2009, Internet release February 2010, http://www.census.gov/population/www/ (accessed March 7, 2010).
19. John F. Gantz, "A Forecast of Worldwide Information Growth through 2010," March 2007, an IDC White Paper, http://www.emc.com/collateral/analyst-reports/expanding-digital-idc-white-paper.pdf (accessed May 1, 2010).
20. John Palfrey and Urs Gasser, *Born Digital: Understanding the First Generation of Digital Natives* (New York: Basic Books, 2008), 1 & 4.

21. Enquiro Research Expert Series, "The Rise of the Digital Native," http://www.enquiro.com/whitepapers/pdf/The-Rise-of-the-Digital-Native.pdf (accessed February 12, 2010).
22. Nicholas Carr, "Is Google Making Us Stupid? What the Internet Is Doing to Our Brains," *Atlantic Monthly,* July/August, 2008, http://www.theatlantic.com/magazine/archive/2008/07/is-google-making-us-stupid/6868/ (accessed May 18, 2010).
23. Carr, "Is Google Making Us Stupid?"
24. Thomas H. Davenport and John C. Beck, *The Attention Economy: Understanding the New Currency of Business* (Boston: Harvard Business School Press, 2001), 2–4.
25. Ed Shane, *Disconnected America: The Consequences of Mass Media in a Narcissistic World* (Armonk, New York: Sharpe, 2001), 40–43.
26. Tyler Cowen, "Three Tweets for the Web," *Wilson Quarterly* 33 (2009): 54–58.
27. National Endowment for the Arts, "Literary Reading in Dramatic Decline," July 8, 2004, http://www.nea.gov/news/news04/ReadingAtRisk.html (accessed May 7, 2010).
28. Joseph Plambeck, "Newspaper Circulation Falls Nearly 9%," *New York Times,* April 27, 2010, B2; Richard Perez-Pena, "The Popular Newsweekly Becomes a Lonely Category," *New York Times,* January 17, 2009, B1 & B2.
29. Jakob Nielsen, *Alertbox,* "F-Shaped Pattern for Reading Web Content," April 17, 2006, http://www.useit.com/alertbox/reading_pattern.html (accessed February 24, 2010).
30. Nielsen, "F-Shaped Pattern."
31. Peter W. Foltz, "Comprehension, Coherence and Strategies in Hypertext and Linear Text," http://www-psych.nmsu.edu/~pfoltz/reprints/Ht-Cognition.html (accessed March 10, 2010).
32. British Library, "Google Generation," http://www.bl.uk/news/pdf/googlegen.pdf, posted January 11, 2008 (accessed March 11, 2010).
33. Carr, "Is Google Making Us Stupid?"
34. "Future of the Internet," http://pewinternet.org/Reports/2010/Future-of-the-Internet-IV.aspx (accessed March 4, 2010).

35. MaryAnn Wolfe, *Proust and the Squid: The Story and Science of the Reading Brain* (New York: Harper Collins, 2007), 214 & 220.
36. Wolfe, 214 & 220.
37. "Mandarin Language Uses More of the Brain than English," June 29, 2003, http://www.futurepundit.com/archives/001427.html (accessed October 21, 2010).
38. Cowen, 54–58.
39. Gary Small and Gigi Vorgan, *iBrain: Surviving the Technological Alteration of the Modern Mind* (New York: Collins Living, 2008), 24–25.
40. Gary Small, et al., "Your Brain on Google: Patterns of Cerebral Activation During Internet Searching," *American Journal of Geriatric Psychiatry*, 17 (2009): 116–26.
41. Cowen, 54–58.
42. Li Shuang, "Scientists Say Speakers of Different Languages Use Separate Parts of Brain," *Global Times,* October 12, 2010, http://www.globaltimes.cn/www/english/sci-edu/china/2010–10/580971.html (accessed October 21, 2010).
43. Christian Gaser and Gottfried Schlaug, "Brain Structures Differ between Musicians and Non-Musicians," *The Journal of Neuroscience* 23, no. 27 (2003, October 8): 9240–245; and Thomas F. Münte, Eckart Altenmüller, and Lutz Jäncke, " The Musician's Brain as a Model of Neuroplasticity," *Nature Reviews Neuroscience* 3 (2002, June): 473–78, Digital Object Identifier No. doi:10.1038/nrn843, http://www.nature.com/nrn/journal/v3/n6/abs/nrn843.html (accessed October 28, 2010).
44. Carr, "Is Google Making Us Stupid?"
45. Cowen, 54–58.
46. Steve Krug, *Don't Make Me Think: A Common Sense Approach to Web Usability* (Berkeley, CA: New Riders, 2006).
47. Susan Jacoby, *The Age of American Unreason* (New York: Pantheon Books, 2008); also, Katherine Washburn and John Thornton, editors, *Dumbing Down: Essays on the Strip Mining of American Culture* (New York: Norton, 1997).

48. Darnton, 76.
49. Carrie B. Fried, "In-Class Laptop Use and Its Effects on Student Learning," *Computers & Education* 50 (April 2008): 906–14.
50. Patricia Cohen, "Fending Off Digital Decay, Bit by Bit," *New York Times,* March 16, 2010, C1 & C6.
51. Mike May, "A Better Lens on Disease," *Scientific American,* May 2010, 74–77; see also Jeffrey M. Perkel, "Digitizing Pathology," *Bioscience Technology* 34 (2010, February 23): 8–12.
52. R. Grant Steen, *The Evolving Brain: The Known and the Unknown* (Amherst, N.Y.: Prometheus, 2007), 361.
53. MacKenzie Smith, "External Bits: How Can We Preserve Digital Files and Save Our Collective Memory?" http://spectrum.ieee.org/computing/hardware/external-bits, posted July 2005 (accessed March 7, 2009); Cohen, 6.
54. Andrew Sullivan, "Why I Blog," *The Atlantic* (2008, November): 106–13.
55. Andrew Keen, *The Cult of the Amateur: How Today's Internet is Killing Our Culture* (New York: Doubleday Currency, 2007), 3, 57, 63.
56. Walter Kirn, "Little Brother Is Watching," *New York Times Sunday Magazine,* October 17, 2010, 17–18; Claire Cain Miller, "The Many Faces of You," *New York Times Week in Review,* October 17, 2010, 2.
57. Rich Barlow, "Learning Not to Share," *Bostonia* (Summer 2010): 16–17.
58. Scott Granneman, "RFID Chips Are Here," June 26, 2003, http://www.securityfocus.com/columnists/169 (accessed March 3, 2010).
59. David Mehegan, "Cursive, Foiled Again," *Boston Globe G Magazine,* January 17, 2009, 14–15; Kitty Burns Florey, *Script and Scribble: The Rise and Fall of Handwriting* (Brooklyn: Melville, 2009).

Chapter 3

1. Gillian Rose, *Visual Technologies* (Thousand Oaks, Calif.: Sage, 2001), 7–8.
2. Nicholas Mirzoeff, *An Introduction to Visual Culture* (New York: Routledge, 1999), 7.

3. National Endowment for the Arts, "Reading at Risk: a Survey of Literary Reading in America, 2004 Report," www.nea.gov (accessed June 7, 2010); and Lionel Shriver, "Missing the Mark," *Wall Street Journal,* October 25, 2008, W3.
4. Quoted by Sue Halpern, "The iPad Revolution," *New York Review of Books,* June 10, 2010, 22.
5. U.S. Department of Education, National Center for Education Statistics, table 271, "Bachelor's Degrees Conferred by Degree-Granting Institutions, by Field of Study: Selected Years, 1970–71 through 2007–08," http://nces.ed.gov/programs/digest/d09/tables/dt09_271.asp?referrer=list (accessed October 29, 2010).
6. John Owens, "The Future of Communications: Film Schools Put More Emphasis on Practical Work," *Chicago Tribune,* June 20, 2004, http://www.google.com/search?sourceid=navclient&ie=UTF-8&rlz=1T4GGLR_enUS310US310&q=increase+number+of+film+schools (accessed October 29, 2010).
7. Nielsen, "Television Audience Report," July 20, 2009, http://blog.nielsen.com/nielsenwire/media_entertainment/more-than-half-the-homes-in-us-have-three-or-more-tvs/ (accessed November 10, 2010).
8. Neal Gabler, "Screaming Extremism," *Boston Globe,* April 24, 2010, A11.
9. Marshall McLuhan, *The Medium Is the Message* (New York: Bantam, 1967).
10. Mizuko Ito et al., "Living and Learning with New Media: Summary of Findings from the Digital Youth Project," John D. and Catherine T. MacArthur Foundation Reports on Digital Media and Learning, posted November 2008, http://digitalyouth.ischool.berkeley.edu/files/report/digitalyouth-WhitePaper.pdf (accessed March 31, 2010).
11. Bob Tedeschi, "Texts from the Lifeguard Chair Are Raising Concerns over Safety," *New York Times,* September 20, 2010, A21.
12. Amanda Lenhart, "Text Messaging Becomes Centerpiece of Communication," April 20, 2010, http://pewresearch.org/pubs/1572/teens-cell-phones-text-messages (accessed April 25, 2010).
13. Lenhart, "Text Messaging."

14. Lenhart, "Text Messaging."
15. Manny Fernandez, "Drama of a Queens Pay Phone," *New York Times,* February 13, 2010, A1 & A3.
16. Damon Darlin, "Behind My Seven Phone Numbers," *New York Sunday Times Business,* September 12, 2010, 8.
17. Matt Richtel, "Hooked on Gadgets, and Paying a Mental Price," *New York Times,* June 7, 2010, A1 & A12.
18. Julie Scelfo, "RU Here, Mom?" *New York Times,* June 10, 2010, D 1.
19. Scelfo, D 7.
20. Ron Kaufman, "Ratings & Advertising 2000: What Are Americans *Really* Watching?!" www.TurnOffYourTV.com, http://www.turnoffyourtv.com/programsratings/ratingsAds.html (accessed May 5, 2010).
21. David Hajdu, "We Are a Camera," *New York Times,* October 15, 2006, sec. 4, 12.
22. Tannis MacBeth Williams, "Background and Overview," in *The Impact of Television: A Natural Experiment in Three Communities,* ed. Tannis MacBeth Williams (Orlando: Fla.: Academic Press, 1986), 18; and Marie Winn, *The Plug-In Drug: Television, Computers and Family Life* (New York: Penguin, 1977, 2002), 8–10.
23. Louise Story, "Anywhere the Eye Can See, It's Likely to See an Ad," *New York Times,* January 15, 2007, http://www.nytimes.com/2007/01/15/business/media/15everywhere.html. (accessed March 31, 2009).
24. Michael Brower and Warren Leon, "Practical Advice from the Union of Concerned Scientists," http://www.ucsusa.org/publications/guide.ch1.html. (accessed March 1, 2010); also bobbie7-ga (Google Answers researcher), "American Advertising in the Media," August, 20, 2002, http://answers.google.com/answers/threadview?id=56750 (accessed March 1, 2010).
25. Guy Trebay, "Whatever Happened to Now?" *New York Times Sunday Magazine,* February 4, 2007, sec. 7, 71–73.
26. Behr Process Corporation, http://www.behr.com/behrx/workbook/index.jsp (April 15, 2008); pamphlets published by Behr Process Corp, 2007, obtained from Home Depot, June 2008.

27. One of the few times was September 20, 1943. Erika Doss, ed., *Looking at Life Magazine* (Washington: Smithsonian Institution Press, 2001), 7.
28. Pamela Paul, *Pornified: How Pornography Is Transforming Our Life, Our Relationships, and Our Families* (New York: Henry Holt, 2005), 53–54.
29. Alessandra Stanley, "Sitcoms' Burden; Too Few Taboos," *New York Times,* September 21, 2008, "Arts and Leisure," 1.
30. Richard Sandomir, "In Bing Crosby's Wine Cellar, Vintage Baseball," *New York Times,* September 24, 2010, 3.
31. Christine Montross, "Dead Body of Knowledge," *New York Times,* March 27, 2009, A23.
32. Harry Hurt, "A Generation with More than Hand-Eye Coordination," *New York Times,* December 21, 2008, B5; Nicholas Mirzoeff, *An Introduction to Visual Culture* (New York: Routledge, 1999), 7.
33. Streeter Seidell, "I Waste People's Time on Line," *New York Times, Week in Review,* April 20, 2008, 2.
34. Seidell, 2; Andrew Keen, *The Cult of the Amateur: How Today's Internet is Killing Our Culture* (New York: Doubleday Currency, 2007), 3.
35. Noam Cohen, "Through Soldiers' Eyes. 'The First YouTube War,'" *New York Times,* May 24, 2010, B3.
36. Bill Pennington, "Colleges' Search for Athletes Continues to Change," *New York Times Sunday Sports,* May 23, 2010, 8.
37. Charles L. Folk, Edward F. Ester, and Kristof Troemel, "How to Keep Attention From Straying: Get Engaged!" *Psychonomic Bulletin & Review* 16 (2009, February): 127–33; Sam Anderson, "In Defense of Distraction," *New York Magazine* 42 (2009, May 25): 28–101.
38. Jenna Wortham, "Staying Informed without Drowning in Data," *New York Times,* December 18, 2008, B8.
39. http://botsblogspresentation.blogspot.com/ (accessed May 20, 2010); cf. also http://technorati.com/blogs/directory/ (accessed May 20, 2010).
40. http://en.wikipedia.org/wiki/Massively_multiplayer_online_role playing_game, (accessed May 20, 2010).

41. http://us.blizzard.com/en-us/company/press/pressreleases.html?081121 (accessed May 20, 2010).
42. http://www.worldofwarcraft.com/info/ and http://en.wikipedia.org/wiki/World_of_Warcraft (accessed May 20, 2010); http://www.gamespot.com/ and http://us.blizzard.com/en-us/company/ (accessed May 20, 2010).
43. http://www.cca-i.com/insight/ and http://www.gbgc.com/services (accessed May 20, 2010).
44. James Paul Gee, *What Video Games Have to Teach Us About Learning and Literacy* (New York: Palgrave, 2003), 5, 8, 23; Hurt, 5.
45. Daniel Solomon, *Global City Blues* (Washington, D.C.: Island Press, 2003), 193.
46. "Brief Biography of Jaron Lanier," http://www.jaronlanier.com/general.html (accessed May 24, 2020).
47. Byron Reeves and J. Leighton Read, *Total Engagement: Using Games and Virtual Worlds to Change the Way People Work and Businesses Compete* (Boston: Harvard Business School Press, 2009).
48. http://www.polhemus.com/ (accessed May 1, 2010).
49. http://whitenoisemp3s.com/free-white-noise (accessed May 15, 2010).
50. "State of the Art of Noise Mapping in Europe," December 2005, http://ec.europa.eu/environment/noise/home.htm; and http://www.londonnoisemap.com/ (accessed March 14, 2010).
51. John Julius Norwich, *A Taste for Travel: An Anthology* (New York: Knopf, 1987); David Lowenthal, *The Past Is a Foreign Country* (New York: Cambridge University Press, 1985), 197.
52. Daniel J. Boorstein, *The Image: A Guide to Pseudo-Events in America* (New York: Vintage Books, 1961, 1987); Neil Postman, *Technopoly: The Surrender of Culture to Technology* (New York: Vintage Books, 1993), 165–66.
53. Nick Paumgarten, "Out to Lunch," *New Yorker,* November 9, 2009, 28–29.
54. Simon Kreckler, Ken Catchpole, and Matthew Bottomley, "Interruptions During Drug Rounds: An Observational Study," *British Journal of Nursing* 17 (2008): 1326–30.

55. Brian Stelter, "The Flipper Challenges the Crawl," *New York Times Week in Review,* December 21, 2008, 3.
56. Postman, *Technopoly*, 135. Neil Postman, *Amusing Ourselves to Death* (New York: Penguin, 1985), 126.
57. Gary Small and Gigi Vorgan, "Meet Your iBrain: How the Technologies That Have Become Part of Our Daily Lives Are Changing the Way We Think," *Scientific American Mind* (November 2008): 43–49.
58. Small and Vorgan, 44.
59. Small and Vorgan, 48.
60. Linda Stone, "The Attention Project," http://lindastone.net/category/attention/continuous-partial-attention/ (accessed February 2, 2010).
61. Gloria Mark, Victor M. Gonzalez, and Justin Harris, "No Task Left Behind? Examining the Nature of Fragmented Work," CHI 2005, *Papers: Take a Number, Stand in Line (Interruptions & Attention 1),* April 2–7, http://www.ics.uci.edu/~gmark/CHI2005.pdf (accessed November 8, 2010).
62. Anderson, 28–101.
63. Anderson, 28–101.
64. A. O. Scott, "The Screening of America," *New York Times Sunday Magazine,* November 23, 2008, 21–22.
65. Brian Stelter, "It's the Show, Not the Screen: Catering to Audiences Who Will Watch a Movie on a Phone," *New York Times,* May 3, 2010, B1 & B3.
66. Virginia Heffernan, "Content and Its Discontent," *New York Times Sunday Magazine,* December 7, 2008, 16–18.
67. Heffernan, 18.
68. Adelbert W. Bronkhorst, "The Cocktail Party Phenomenon: A Review on Speech Intelligibility in Multiple-Talker Conditions," *Acta Acustica United with Acustica* 86 (2000): 117–28, http://eaa-fenestra.org/products/acta-acustica/most-cited/acta_86_2000_Bronkhorst.pdf (accessed March 31, 2010). See also E. C. Cherry, "Some Experiments on the Recognition of Speech, with One and with Two Ears," *Journal of Acoustical Society of America* 25 (1953): 975–79.

69. Rhonda Martinussen et al., "Bottom of Form: A Meta-Analysis of Working Memory Impairments in Children with Attention-Deficit/Hyperactivity Disorder," *Journal of the American Academy of Child and Adolescent Psychiatry* 44, no. 4 (April 2005): 377–84; and Margaret A. Sheridan, Stephen Hinshaw, and Mark D'Esposito, "Efficiency of the Prefrontal Cortex During Working Memory in Attention-Deficit/Hyperactivity Disorder," *Journal of the American Academy of Child and Adolescent Psychiatry* 46, no. 10 (October, 2007): 1357–66.
70. Michael Streich, "American Students and the Decline in History," February 27, 2009, http://www.suite101.com/content/american-students-and-the-decline-in-history-a99242 (accessed November 11, 2010); and NBC Los Angeles, "Study: Americans Don't Know Much About History," January 26, 2009, http://www.nbclosangeles.com/news/local-beat/Study-Americans-Dont-Know-About-Much-About-History.html (accessed November 11, 2010).
71. Sebastian Smee, "Science, Art Share Aerial Imagery's Goals," *Boston Globe,* October 13, 2010, B2.

Chapter 4

1. Marlene A. Lee and Mark Mather, "U.S. Labor Force Trends," *Population Bulletin* 63, no. 2 (2008): 7, http://www.prb.org/pdf08/63.2uslabor.pdf (accessed September 29, 2010); "National Economic Conditions and Trends," http://www.ers.usda.gov/Publications/rct71/rct71c.pdf (accessed June 8, 2010).
2. Dean Horvath, "Knowledge Worker," February 24, 1999, http://searchcrm.techtarget.com/definition/knowledge-worker (accessed June 09, 2010).
3. Peter Drucker, *Landmarks of Tomorrow* (New York: Harper, 1959).
4. U.S. Bureau of Labor Statistics, "Occupational Employment and Wages News Release," May 14, 2010, http://www.bls.gov/news.release/ocwage.htm (accessed June 2, 2010).
5. Massachusetts Institute of Technology, Sloan School of Management, "Understanding Productivity in the Information Age," http://mitsloan.mit.edu/newsroom/2007-brynjolfsson.php (accessed June 28, 2010).

6. Jack E. Triplett, "The Solow Productivity Paradox: What Do Computers Do to Productivity?" *Canadian Journal of Economics (Revue Canadienne d'Economique)* 32 (March 2, 1999), http://www.brookings.edu/~/media/Files/rc/articles/1999/04technology_triplett02/199904.pdf (accessed June 9, 2010). See also Erik Brynjolfsson and Lorin Hitt, "Beyond the Productivity Paradox: Computers Are The Catalyst For Bigger Changes," *Communications of the ACM* 41, no. 8 (August 1998): 49–55.
7. Catherine Rampell, "In a Job Market Realignment, Some Workers No Longer Fit," *New York Times,* May 13, 2010, 1 & B12.
8. U.S. Bureau of Labor Statistics, *Standard Occupation Classification,* http://www.bls.gov/soc/#classification (accessed June 6, 2010).
9. U.S. Bureau of Labor Statistics, "Number of Jobs Held, Labor Market Activity, and Earnings Growth among the Youngest Baby Boomers: Results from a Longitudinal Survey," June, 2008, www.bls.gov/news.release/pdf/nlsoy.pdf (accessed June 28, 2010).
10. Patrick Dawson, *Reshaping Change: A Processual Approach to Understanding Change* (London: Routledge, 2003), 2.
11. Debra E. Meyerson, "Radical Change, The Quiet Way," *Harvard Business Review On Culture and Change* (Boston: Harvard Business School Press, 2006), 64. Originally published October 2001, reprint R0109F.
12. Paul Strebel, "Why Do Employees Resist Change?" *Harvard Business Review On Culture and Change* (Boston: Harvard Business School Press, 2009), 139. Originally published May-June 1996, reprint 96310.
13. Jim Collins, "The Subject of Enduring Greatness," *Fortune,* May 5, 2008, 74.
14. Rodney Carlisle, *Scientific American: Inventions and Discoveries* (Hoboken, N.J.: Wiley, 2004), 414.
15. Joseph A. Schumpeter, *Capitalism, Socialism and Democracy* (New York: Harper, 1975) [orig. pub. 1942], 82–85.
16. Shumpeter, *Capitalism,* 84.
17. U.S. Department of Labor, Bureau of Labor Statistics, "Employment and Earnings, 2009," November 2009, http://www.dol.gov/wb/stats/main.htm (accessed June 25, 2010); National Bureau of Labor Statistics,

"Women Make Up . . . Half of the Workforce," Friday, October 8, 2010, http://www.bls.gov/news.release/archives/empsit_10082010.pdf (accessed November 3, 2010). Also, Joint Economic Committee, U.S. Congress, "Women and the Economy 2010," August 26, 2010, http://jec.senate.gov/public/?a=Files.Serve&File_id=8be22cb0–8ed0–4a1a-841b-aa91dc55fa81 (accessed November 2, 2010).

18. Julie Bosman, "A Princeton Maverick Succumbs to a Cultural Shift," *New York Times,* January 3, 2007, B7.
19. Jacques Barzun, *From Dawn to Decadence* (New York: HarperCollins, 2001), 784.
20. Lee and Mather, 1.
21. James Fallow, *Postcards from Tomorrow Square: Reports from China* (New York: Vintage Books, 2009), 96.
22. Peggy Holman and Tom Devane, "Introduction: The Changing Nature of Change," in *Change Handbook: Group Methods for Shaping the Future,* eds. Peggy Holman and Tom Devane (San Francisco: Berrett-Koehler Publications, 1999), 2.
23. Ross Kerber, "Withdrawing from the ATM Habit," *Boston Globe,* February 19, 2008, http://www.boston.com/business/personalfinance/articles/2008/02/19/withdrawing_from_the_atm_habit/ (accessed June 8, 2010).
24. Personal interview, June 24, 2010.
25. Jeremy Simon, "Origin of Credit Cards," posted February 7, 2008, http://www.creditcards.com/credit-card-news/origin-of-credit-cards-1276.php (accessed June 8, 2010).
26. Kerber, "Withdrawing from the ATM Habit."
27. Holman and Devane, "Introduction: The Changing Nature of Change," 2.
28. U.S. Census Bureau, *E-Stats,* http://www.census.gov/retail/mrts/www/data/html/10Q1table1.html (accessed June 15, 2010).
29. U.S. Census Bureau, *E-Stats,* http://www.census.gov/econ/estats/2008/2008reportfinal.pdf (accessed June 14, 2010).
30. Nielsen Online, http://www.nielsen-online.com/pr/pr_081112.pdf (accessed June 14, 2010).

31. Victor M. González and Gloria Mark, "'Constant, Constant, Multi-tasking Craziness': Managing Multiple Working Spheres," *Proceedings of ACM CHI 2004*, April 2004, 113–20, http://www.ics.uci.edu/~gmark/CHI2004.pdf (accessed June 16, 2010).
32. Gloria Mark, Victor M. González, and Justin Harris, "No Task Left Behind? Examining the Nature of Fragmented Work," *Proceedings of SIGCHI Conference on Human Factors in Computing Systems 2005,* 321–30, http://www.ics.uci.edu/~gmark/ (accessed February 22, 2010).
33. Jon Hamilton, "Think You're Multitasking? Think Again," National Public Radio, *Morning Edition,* October 2, 2008, http://www.npr.org/templates/story/story.php?storyId=95256794 (accessed February 15, 2010).
34. Gloria Mark, Daniela Gudith, and Ulrich Klocke, "The Cost of Interrupted Work: More Speed and Stress," *Proceedings of SIGCHI Conference on Human Factors in Computing Systems 2008*, http://www.ics.uci.edu/~gmark/chi08-mark.pdf (accessed June15, 2010). I follow Mark, Gudith, and Kloche closely in these paragraphs.
35. Mark, González, and Harris, 324.
36. Vince Poscente, *The Age of Speed: Learning to Thrive in a More-Faster-Now World* (Austin, Texas: Bard Press, 2008), 8.
37. "Sherry Turkle Interview," *Frontline: Digital Nation,* Public Broadcasting Service, February 2, 2010, www.pbs.org/wgbh/pages/frontline/digitalnation/interviews/turkle.html/ (accessed February 24, 2010).
38. Sylvain Charron and Etienne Koechlin, "Divided Representation of Concurrent Goals in the Human Frontal Lobes," *Science* 328, no. 5976 (April 16, 2010): 360–63, Digital Object Identifier No. doi: 10.1126/science.1183614, http://www.sciencemag.org/cgi/content/short/328/5976/360 (accessed June 18, 2010).
39. Jon Hamilton, "Multitasking Brain Divides and Conquers, to a Point," NPR, *All Things Considered,* April 15, 2010, http://www.npr.org/templates/story/story.php?storyId=126018694 (accessed June 18, 2010).
40. Katherine Harmon, "Motivated Multitasking: How the Brain Keeps Tabs on Two Tasks at Once," *Scientific American,* April 15, 2010,

http://www.scientificamerican.com/article.cfm?id=multitasking-two-tasks (accessed June 18, 2010).

41. Jon Hamilton, "Multitasking in the Car: Just Like Drunken Driving," NPR, *Morning Edition,* October 16, 2008, http://www.npr.org/templates/story/story.php?storyId=95784052&ps=rs (accessed June 18, 2010).
42. Matt Richel, "Hooked on Gadgets and Paying a Mental Price," *New York Times,* June 7, 2010, A12.
43. Paul Raeburn, "Multitasking May Not Mean Higher Productivity," NPR, *Science Friday,* August 28, 2009, http://www.npr.org/templates/story/story.php?storyId=112334449&ps=rs (accessed June 18, 2010).
44. Eyal Ophir, Clifford Nass, and Anthony D. Wagner, "Cognitive Control in Media Multitaskers," *Proceedings of the National Academy of Sciences,* July 20, 2009, 1–5, http://www.pnas.org/content/early/2009/08/21/0903620106.full.pdf+html (accessed June 19, 2010).
45. Peter Borsay, *A History of Leisure* (New York: Palgrave Macmillan, 2006), 2–3.
46. U.S. Bureau of Labor Statistics, "Occupational Employment and Wages News Release," May 14, 2010, http://www.bls.gov/news.release/ocwage.htm (accessed June 1, 2010).
47. Alison Lobron, "Is Unlimited Vacation a Good Thing?" *Boston Sunday Globe Magazine,* July 20, 2008, 8.
48. The International Health, Racquet, and Sportsclub Association, *Survey of International Health, Racquet and Sportsclub Association Members,* figures for 2010, http://live.ihrsa.org/index.cfm?fuseaction=Page.viewPage&pageId=18735&parentID=18745&grandparentID=18275&nodeID=15 (accessed June 27, 2010).
49. Linda Nazareth, *The Leisure Economy: How Changing Demographics, Economics, and Generational Attitudes will Reshape Our lives and Our Industries* (Ontario: Wiley, 2007), 12–13.
50. U.S. Bureau of Labor Statistics, Table 11, "Time Spent in Leisure and Sports Activities for the Civilian Population by Selected Characteristics," *American Time Use Survey Summary,* June 24, 2009, http://www.bls.gov/news.release/atus.nr0.htm (accessed June 2, 2010).

51. Arlie Russell Hochschild, *The Time Bind: When Work Becomes Home and Home Becomes Work* (New York: Henry Holt, 1997).
52. Jerry A. Jacobs and Kathleen Gerson, "Understanding Changes in American Working Time: A Synthesis," in Cynthia F. Epstein and Arne L Kalleberg, eds., *Fighting for Time: Shifting Boundaries of Work and Social Life* (New York: Russell Sage Foundation, 2004), 29.
53. Jacobs and Gerson, 18–20.
54. Ronald R. Sims, "General Introduction and Overview of the Book," in Ronald R. Sims, ed., *Changing the Way We Manage Change* (Westport, CT: Quorum Books, 2002), 1.

Chapter 5

1. Jane E. Brody, "Turning the Ride to School into a Walk," *New York Times,* September 11, 2007, D7.
2. Lenore Skenazy, *Free-Range Kids: Giving Our Children the Freedom We Had without Going Nuts with Worry* (San Francisco: Jossey-Bass, 2009), 192–93.
3. Helena Echlin, "Is It Rude to Eat on the Subway?" July 15, 2008, http://www.chow.com/stories/11211 (accessed July 6, 2010). See also, "Is It Poor Etiquette to Eat in Public around People Who Aren't Eating?" at http://ca.answers.yahoo.com/question/index?qid=20100607005706AAROxj3 (accessed July 6, 2010).
4. Michael Pollan, "The Food Movement, Rising," *New York Review of Books,* June 10, 2010, 31–33.
5. Pollan, 31.
6. Rand Richards Cooper, "It's My Party, and You Have to Answer," published March 14, 2010, http://www.nytimes.com/2010/03/15/opinion/15cooper.html?_r=1 (accessed July 7, 2010).
7. International Telecommunication Union, "Number of Cell Phones Worldwide Hits 4.6B," February 15, 2010, http://www.cbsnews.com/stories/2010/02/15/business/main6209772.shtml (accessed July 8, 2010).
8. Daniel E. Sullivan, "Recycled Cell Phones—A Treasure Trove of Valuable Metals," U.S. Geological Survey Fact Sheet 2006, 3097, http://pubs.usgs.gov/fs/2006/3097/ (accessed July 8, 2010).

9. Telephony Museum, "History of the Phone," http://www.telephony museum.com/History%201940-today.htm (accessed July 8, 2010).
10. Christine Pearson, "Sending a Message That You Don't Care," *New York Times Sunday Business,* May 16, 2010, 9.
11. Pearson, "Sending a Message," 9.
12. Douglas Stewart, "Statistics and Cell Phones," posted August 27, 2008, http://www.articlesbase.com/computers-articles/statistics-and-cell-phones-538109.html (accessed July 16, 2010).
13. K. Lynn, "Cell Phone Addiction?" May 12, 2008, http://www.arti clesbase.com/computers-articles/cell-phone-addiction-411578.html (accessed June 11, 2010).
14. Personal e-mail to author from Dr. Jacqueline Alfonso, addiction specialist, April 7, 2010.
15. See for example, Ashley Halsey III, "U.S. Bans Truckers, Bus Drivers From Texting While Driving," *Washington Post,* January 27, 2010, http://www.washingtonpost.com/wp-dyn/content/article/2010/01/26/AR2010012602031.html (accessed July 10, 2010); and Michael Levenson, "State Inches Ahead on Banning Texting While Driving," *Boston Globe,* January 29, 2010, http://www.boston.com/news/local/massachusetts/articles/2010/01/29/state_inches_ahead_on_banning_texting_while_driving/ (accessed July 10, 2010).
16. W3.org, "Jargon and Terms," http://www.w3.org/Terms.html (accessed October 23, 2008); and David Crystal, *txtng: the gr8 db8* (Oxford: Oxford University Press, 2008), 8.
17. Kate Jackson, *Mean and Lowly Things* (Cambridge: Harvard University Press, 2008), 117, 180.
18. Naomi S. Baron, *Always On: Language in an Online and Mobile World* (New York: Oxford University Press, 2008), 28.
19. Eric Moskowitz, "Sexting Probed as Child Porn: Nude Photo of Girl Reportedly Sent," *Boston Globe,* March 6, 2010, http://www.boston.com/news/local/massachusetts/articles/2010/03/06/sexting_probed_as_child_porn/ (accessed July 7, 2010); also "'Sexting' Shockingly Common Among Teens," January, 15, 2009, http://www.cbs news.com/stories/2009/01/15/national/main4723161.shtml (accessed July 7, 2010).

20. National Campaign to Prevent Teen and Unplanned Pregnancy, "Sex and Tech: Results from a Survey of Teens and Young Adults," http://www.thenationalcampaign.org/sextech/PDF/SexTech_Summary.pdf (accessed July 8, 2010).
21. Katherine M. Flegal, Margaret D. Carroll, Cynthia L. Ogden, and Lester R. Curtin, "Prevalence and Trends in Obesity Among U.S. Adults, 1999–2008," *Journal of the American Medical Association* 303, no. 3 (2010): 235–41. Published online January 13, 2010 (doi:10.1001/jama.2009.2014), http://jama.ama-assn.org/cgi/content/full/303/3/235?ijkey=ijKHq6YbJn3Oo&keytype=ref&siteid=amajnl (accessed July 10, 2010).
22. John Jervis, *Exploring the Modern: Patterns of Western Culture and Civilization* (Malden, Mass.: Blackwell, 1998), 127–28.
23. David Colman, "After Years of Being Out, the Necktie Is In," *New York Times,* Thursday, October 11, 2007, E1 & E5.
24. Niko Koppel, "Are Your Jeans Sagging? Go Directly to Jail," *New York Times,* Thursday, August 30, 2007, E1 & E8.
25. Rob Walker, "Sanctioned Subversion," *New York Times Sunday Magazine,* May 2, 2010, 24.
26. Snopes. com, "Hat Trick," http://www.snopes.com/history/american/jfkhat.asp, September 27, 2007 (accessed May 10, 2010).
27. VanishingTattoo.com, "Tattoo Facts and Statistics," http://www.vanishingtattoo.com/tattoo_facts.htm (accessed July 10, 2010).
28. Pew Research Center, "A Portrait of 'Generation Next': How Young People View Their Lives, Futures and Politics," January 9, 2007, http://people-press.org/report/300/a-portrait-of-generation-next (accessed July 10, 2010).
29. Tamar Lewin, "In Twist for High School Wrestlers, Girl Flips Boy," *New York Times,* Saturday, February 17, 2007, A1 & A13.
30. Michael Wilson, "A Fixture that Became a Rarity: Cigarette Machines Go the Way of Smoke," *New York Times,* July 12, 2010, A18.
31. Ellen Gemerman, "Are Misbehavin': No Tonys for These Performances," *Wall Street Journal,* June 6, 2009, A1 & A4.

32. Snopes.com, "The Shirt Off His Back," http://www.snopes.com/movies/actors/gable1.asp, posted August 9, 2007 (accessed July 13, 2010).
33. "Angelina Jolie Tattoos," http://www.freetattoodesigns.org/angelina-jolie-tattoos.html (accessed July 13, 2010).
34. "Lady Gaga Explains the Meaning behind Her New Tattoos," October 11, 2009, http://anythinghollywood.com/2009/10/lady-gaga-explains-the-meaning-behind-her-new-tattoos/ (accessed July 13, 2010).
35. "Lady Gaga Explains."
36. Everett M. Rogers, *Diffusion of Innovations,* 5th ed. (New York: Free Press, 2003), 5.
37. Rogers, 283–84.
38. Rogers, 23, 31.
39. WAC Survey and Strategic Consulting, "Research on Trends: Influentials, Innovators & Early Adopters," February 23, 2010, http://www.greenbook.org/marketing-research.cfm/influentials-innovators-early-adopters (accessed July 13, 2010).
40. Rogers, 103.
41. Robin Abrahams, *Miss Conduct's Mind over Manners: Master the Slippery Rules of Modern Ethics and Etiquette* (New York: Henry Holt, 2009), 1.
42. Bill Bishop, *The Big Sort: Why the Clustering of Like-Minded America Is Tearing Us Apart* (Boston: Houghton Mifflin, 2008).
43. Abrahams, 2–3.
44. Steve Farkas, Jean Johnson, Ann Duffett, and Kathleen Collins, "Aggravating Circumstances: A Status Report on Rudeness in America," December 31, 2001, http://www.publicagenda.org/reports/aggravating-circumstances, 31 (accessed July 13, 2010).
45. Farkas, Johnson, Duffett, and Collins, "Aggravating Circumstances."

Chapter 6

1. Stephen Kern, *The Culture of Time and Space: 1880–1918* (Cambridge, Mass.: Harvard University Press, 1983), 129.

2. Thomas Hylland Eriksen, *Tyranny of the Moment: Fast and Slow Time in the Information Age* (Sterling, Va.: Pluto Press, 2001), 102.
3. Eriksen, 101–2.
4. Eriksen, 85.
5. John P. Robinson and Geoffrey Godbey, *Time for Life: The Surprising Ways Americans Use Their Time* (University Park: Pennsylvania State University, 1997), 136–37.
6. Robinson and Godbey, 137.
7. Kevin Sites, *In the Hot Zone* (New York: Harper Perennial, 2007), 12.
8. Jervis, 208.
9. Eriksen, 38.
10. Kern, 110–11.
11. Kern, 10–12.
12. Jervis, 258.
13. Jervis, 211.
14. Jervis, 212.
15. Kern, 314–16.
16. Robert Levine, *A Geography of Time* (New York: Basic Books, 1997), 9.
17. Richard Wiseman, "Pace of Life," http://www.paceoflife.co.uk/ (accessed July 5, 2010).
18. British Broadcasting Corporation, "What Walking Speeds Say About Us," http://news.bbc.co.uk/1/hi/magazine/6614637.stm (accessed September 18, 2007).
19. James Gleick, *Faster: The Acceleration of Just About Everything* (New York: Pantheon Books,1999), 6.
20. Gleick, 7.
21. "Collection of Estimates of the Quantities of Data Contained from Various Media," http://www.uplink.freeuk.com/data.html (accessed July 27, 2010).
22. Thom Hickey, "Entire Library of Congress," on *Outgoing: Library Metadata Techniques and Trends,* June 21, 2005, http://outgoing.typepad.com/outgoing/2005/06/entire_library_.html (accessed July 27, 2010).

23. "Gigabyte," www.Webopedia.com (accessed August 1, 2010).
24. "World Network Speed Record Quadrupled," December 2, 2004, http://www.ir35calc.co.uk/world_network_speed_record.aspx (accessed June 3, 2010).
25. Apple iPhone, "iPhone 4 Technical Specifications," http://www.apple.com/iphone/specs.html (accessed July 27, 2010).
26. Gleick, 65–67.
27. Arlie Russell Hochschild, *The Time Bind: When Work Becomes Home and Home Becomes Work* (New York: Henry Holt, 1997), 13.
28. Jerry A. Jacobs and Kathleen Gerson, "Understanding Changes in American Working Time: A Synthesis," in *Fighting for Time: Shifting Boundaries of Work and Social Life,* eds. Cynthia F. Epstein and Arne L Kalleberg (New York: Russell Sage Foundation, 2004), 26.
29. Jacobs and Gerson, 33.
30. Linda Nazareth, *The Leisure Economy: How Changing Demographics, Economics, and Generational Attitudes Will Reshape Our Lives and Our Industries* (Ontario: Wiley, 2007), 15.
31. Edward Cornish, "Television: The Great Time-Eating Machine," book review of John P. Robinson and Geoffrey Godbey, *Time for Life: The Surprising Ways Americans Use Their Time,* in *Futurist* 32, no. 1, (1998): 60.
32. John P. Robinson and Geoffrey Godbey, *Time for Life: The Surprising Ways Americans Use Their Time* (University Park: State University of Pennsylvania, 1997), 293.
33. Hal R. Varian, "Still Waiting for the Future of Leisure to Arrive," *New York Times,* March 8, 2007, C3.
34. U.S. Bureau of Labor Statistics, "American Time Use Survey Summary," June 22, 2010, http://www.bls.gov/news.release/atus.nr0.htm (accessed June 29, 2010); Robinson and Godbey, xvii, 148.
35. Robinson and Godbey, 294.
36. Victoria J. Rideout, Ulla G. Foehr, and Donald F. Roberts, "Generation M2: Media in the Lives of 8–18-year Olds," Kaiser Family Foundation Study, January 2010, http://www.kff.org/entmedia/upload/8010.pdf (accessed August 16, 2010).
37. Rideout, Feohr, and Roberts, "Generation M2," 2.

38. Amanda Lenhart, "Text Messaging Becomes Centerpiece Communication," Pew Internet & American Life Project, April 20, 2010, http://pewresearch.org/pubs/1572/teens-cell-phones-text-messages (accessed August 16, 2010).
39. Michael Malone, "TVB Study: Adults Spend Twice as Much Time on TV Than Web," May 25, 2010, http://www.broadcastingcable.com/article/453033-TVB_Study_Adults_Spend_Twice_as_Much_Time_on_TV_Than_Web.php (accessed August 5, 2010).
40. Gleick, 12.
41. Eriksen, 70.
42. Eriksen, 59.
43. Paul H. Rubin, "Instant Info a Two-Edged Sword: The Internet Facilitates Markets, and Bubbles," December 31, 2008, http://online.wsj.com/article/SB123068195686844065.html?KEYWORDS=Instant+information+is+a+two-edged+sword (accessed August 19, 2010).
44. Jonathan Abrams and John Branch, "Fast and Risky, Sledding Track Drew Red Flags," *New York Times,* February 14, 2010, 1 & 4.
45. Anand Giridharadas, "Getting in (and Out of) Line," *Sunday New York Times, Week in Review,* August 8, 2010, 5.
46. Giridharadas, 5.
47. Neil Gabler, "Impatient for Change: In a Society Where Everything Is Instant, Government Stays Slow," *Boston Globe,* July 6, 2010, A11.
48. Bodil Jonsson, *Unwinding the Clock* (New York: Harcourt, 1999), 2 & 17.
49. Jervis, 323.
50. Morris Berman, *The Twilight of American Culture* (New York: Norton, 2000), 49.
51. David Michael Bruno, "100 Thing Challenge," posted August 2010, http://www.guynameddave.com/100-thing-challenge.html (accessed August 15, 2010).
52. Stephanie Rosenbloom, "But Will It Make You Happy?" *Sunday New York Times Business Section,* August 8, 2010, 1 & 4.
53. For example, Ronni Eisenberg, *Organize Your Life: Free Yourself from Clutter and Find More Personal Time* (Hoboken, NJ: Wiley, 2007);

Aimee Baldridge, *Organize Your Digital Life* (Washington, DC: National Geographic, 2009); Joyce Meyer, *100 Ways to Simplify Your Life* (New York: Faith Words, 2008); Marcia Ramsland, *Simplify Your Space: Create Order and Reduce Stress* (Nashville, TN: Thomas Nelson, 2007); Werner Küstenmacher, *How to Simplify Your Life: Seven Practical Steps to Letting Go of Your Burdens and Living a Happier Life* (New York; McGraw-Hill, 2004).

54. National Sleep Foundation, "One-Third of Americans Lose Sleep Over Economy," March 1, 2009, http://www.sleepfoundation.org/article/press-release/one-third-americans-lose-sleep-over-economy (accessed August 19, 2010).
55. Ibid.
56. Matt Richtel, "Outdoors and Out of Reach: Studying the Brain," *New York Times,* August 16, 2010, A1 & A10.
57. Joseph Pieper, *Leisure, the Basis of Culture* (original edition, Pantheon Books, 1952). Translation by Gerald Malsbary (South Bend: St. Augustine's Press, 1998), 20.
58. Pieper, 25.
59. Pieper, 29.

Chapter 7

1. Paul R. Amato et al., *Alone Together: How Marriage in America Is Changing* (Cambridge, MA: Harvard University Press, 2007), 11.
2. Amato, *Alone Together,* 1.
3. U.S. Census Bureau, "Households by Type and Size: 1900 to 2002," http://www.census.gov/statab/hist/HS-12.pdf (accessed September 1, 2010); and U.S. Census Bureau, "2006–2008 American Community Survey 3-Year Estimates," http://factfinder.census.gov/servlet/ACSSAFFFacts (accessed August 23, 2010).
4. Tommie J. Hamner and Pauline H. Turner, *Parenting in Contemporary Society* (Boston: Allyn and Bacon, 2001), 2.
5. P. Y. Goodwin, W. D. Mosher, and A. Chandra, "Marriage and Cohabitation in the United States: A Statistical Portrait Based on Cycle 6 (2002) of the National Survey of Family Growth," *Vital*

Health Statistics 23, no. 28, posted 2010, http://www.cdc.gov/nchs/data/series/sr_23/sr23_028.pdf (accessed August 23, 2010).

6. U.S. Census Bureau, "Census Bureau Reports Families With Children Increasingly Face Unemployment," January 15, 2010, http://www.census.gov/newsroom/releases/archives/families_households/cb10–08.html (accessed September 1, 2010).
7. Goodwin, Mosher, and Chandra, "Marriage and Cohabitation," 10.
8. Paul R. Amato, "Recent Changes in Family Structure: Implications for Children, Adults, and Society," commissioned by the National Healthy Marriage Resource Center, April 2008, 1–6, http://www.healthymarriageinfo.org/docs/changefamstructure.pdf (accessed August 23, 2010).
9. Joyce A. Martin, B. E. Hamilton, P. D. Sutton, S. J. Ventura, et al., "Births: Final Data for 2006," *National Vital Statistics Reports* 57, no. 7 (2009, January 7): http://www.cdc.gov/nchs/data/nvsr/nvsr57/nvsr57_07.pdf (accessed September 1, 2010), 11.
10. Goodwin, Mosher, and Chandra, "Marriage and Cohabitation," 4.
11. Centers for Disease Control and Prevention, "About Teen Pregnancy, An Update," 2009, http://www.cdc.gov/reproductivehealth/adolescentreprohealth/AboutTP.htm (accessed August 23, 2010).
12. Martin et al., "Births," 12.
13. Paul R. Amato, "Recent Changes in Family Structure," p. 6.
14. Amato, "Recent Changes," 4–6.
15. U.S. Centers for Disease Control and Prevention, National Center for Health Statistics, "National Survey of Family Growth Statistics for 2002," May 2008, http://www.cdc.gov/nchs/data/infosheets/infosheet_nsfg.htm (accessed September 7, 2010).
16. Michael Pollan, "The Food Movement, Rising," *New York Review of Books,* June 10, 2010, 33.
17. Thomas Hylland Eriksen, *Tyranny of the Moment: Fast and Slow Time in the Information Age* (Sterling, VA: Pluto Press, 2001), 132–33.
18. U.S. Bureau of Labor Statistics, "Labor Force Statistics from the Current Population Survey: Women in the Labor Force: A Databook" (2009 edition), http://www.bls.gov/cps/wlf-intro-2009.htm (accessed August 24,

2010); and Catherine Rampell, "Women Now a Majority in American Workplaces," February 5, 2010, http://www.nytimes.com/2010/02/06/business/economy/06women.html?_r=1&ref=business (accessed August 24, 2010).

19. Marlene A. Lee and Mark Mather, "U.S. Labor Force Trends," *Population Bulletin* 63, no. 2 (2008): 4, http://www.prb.org/pdf08/63.2uslabor.pdf (accessed September 29, 2010).
20. Hanna Rosin, "The End of Men," *Atlantic* (2010, July/August): 58, 60.
21. Rosin, "The End of Men," 60, 62; U.S. Bureau of Labor Statistics, "Labor Force Statistics from the Current Population Survey: Women in the Labor Force: A Databook."
22. David Leonhart, "A Market Punishing to Mothers," *New York Times,* August 4, 2010, B1 & 6.
23. TitleIX.info, "History of Title IX," http://www.titleix.info/History/History-Overview.aspx (accessed September 6, 2010); and U.S. Department of Labor, "Title IX, Education Amendments of 1972," http://www.dol.gov/oasam/regs/statutes/titleix.htm (accessed September 6, 2010).
24. American Association of University Women, "Title IX Statistics," http://www.aauw.org/act/laf/library/athleticStatistics.cfm (accessed September 6, 2010).
25. Olympic.org, http://www.olympic.org/en/content/Sports/ (accessed September 6, 2010); and Independent Women's Football League, "IWFL about the IWFL," http://www.iwflsports.com/abouttheiwfl.php (accessed September 6, 2010).
26. U.S. Census Bureau, "Census Bureau Reports Nearly 6 in 10 Advanced Degree Holders Age 25–29 Are Women," April 20, 2010, http://www.census.gov/newsroom/releases/archives/education/cb10–55.html (accessed August 23, 2010).
27. Jennifer Delahunty Britz, "To All the Girls I've Rejected," *New York Times,* March 23, 2006, http://www.nytimes.com/2006/03/23/opinion/23britz.html (accessed August 24, 2010).
28. Rosin, "The End of Men," 67–68.
29. Amato, *Alone Together,* 235.

30. U.S. Census Bureau, Facts for Features, "Valentine's Day 2009: February 14," http://www.census.gov/newsroom/releases/archives/facts_for_features_special_editions/cb09-ff02.html (accessed September 1, 2010).
31. See ConsumerRankings.com, "Top 5 Online Dating Sites of 2010," http://www.consumer-rankings.com/Dating/index-Dating.aspx (accessed August 30, 2010).
32. See, for instance, Natalie Southwick, "The Bride Wore Flip-Flops," *Boston Globe,* September 4, 2010, Section G, 12–13.
33. See, for example, http://www.youtube.com/results?search_query=outrageous+weddings&aq=0 (accessed September 15, 2010).
34. US Marriage Laws, http://www.usmarriagelaws.com/search/united_states/officiants_requirements/index.shtml (accessed August 24, 2010).
35. "Market Totals: Estimated Weddings," http://www.theweddingreport.com/wmdb/index.cfm?action=db.viewdetail&brand=google&gclid=CP3amLeC4qMCFeoD5QodzVZ4nA (accessed August 27, 2010); and Ellen Terrell, "Wedding Industry Research," 2004, http://www.loc.gov/rr/business/wedding (accessed August 27, 2010).
36. Gay Marriage Facts, http://www.gaymarriagefacts.net/ (accessed, August 15, 2010).
37. Amato, *Alone Together,* 12, 31.
38. "Births, Marriages, Divorces, and Deaths: Provisional Data for 2008," *National Vital Statistics Reports* 57, no. 19 (July 29, 2009): http://www.cdc.gov/nchs/data/nvsr/nvsr57/nvsr57_19.pdf (accessed September 9, 2010), 1; and Amato, "Recent Changes," 4–6.
39. Amato, "Recent Changes," 4–5.
40. Pew Research Center Publications, "Social Isolation and New Technology: How the Internet and Mobile Phones Impact Americans' Social Networks," November 4, 2009, http://pewresearch.org/pubs/1398/internet-mobile-phones-impact-american-social-networks (accessed September 9, 2010); and Veronica M. Scott, Karen E. Mottarella, and Maria J. Lavooy, "Does Virtual Intimacy Exist? A Brief Exploration into Reported Levels of Intimacy in Online Relationships,"

CyberPsychology & Behavior 9, no. 6 (December 2006): 759–61, Digital Object Identifier No. doi: 10.1089/cpb.2006.9.759 (accessed September 8, 2010).

41. Miller McPherson, Lynn Smith-Lovin, and Mathew E. Brashears, "Social Isolation in America: Changes in Core Discussion Networks over Two Decades," *American Sociological Review* 71, no. 3 (June 2006): 353–75.
42 Arno, *Alone Together,* 207, 236.
43. Abigail Sullivan Moore, "Failure to Communicate," *New York Sunday Times Education Life* (Special Section), July 25, 2010, 20–21.
44. Moore, "Failure to Communicate," 20–21.
45. Moore, 20–21.
46. Trip Gabriel, "Students, Welcome to College; Parents, Go Home," *New York Times,* August 23, 2010, A1 & A3.
47. Jennifer Ludden, "Teen Texting Soars; Will Social Skills Suffer?" National Public Radio broadcast, April 20, 2010, http://www.npr.org/templates/story/story.php?storyId=126117811&ps=cprs (accessed August 31, 2010).
48. Robert D. Putnam, *Bowling Alone: The Collapse and Revival of American Community* (New York: Simon & Schuster, 2000).
49. Claude S. Fischer, "*Bowling Alone:* What's the Score?" Paper to be presented to the "Author Meets Critic: Putnam, *Bowling Alone*" session of the meetings of the American Sociological Association, Anaheim, California, August 2001, http://ucdata.berkeley.edu/rsfcensus/papers/BowlingAlone.pdf (accessed August 26, 2010).
50. Jeffrey Jensen Arnett, "Emerging Adulthood: A Theory of Development from the Late Teens through the Twenties," *American Psychologist* 55, no. 5 (2000): 477.
51. Arnett, "Emerging Adulthood," 478.
52. National Center for Health Statistics, "Information Sheet," May 2009, National Vital Statistics System, http://www.cdc.gov/nchs/data/infosheets/infosheet_NVSS.htm (accessed September 1, 2010).
53. E–mail to author from Kimberley B. Balkus, Director, MIT Alumni Records, October 20, 2010.

Chapter 8

1. U.S. Census Bureau, "Income, Poverty and Health Insurance Coverage in the United States: 2009," September 16, 2010, http://www.census.gov/newsroom/releases/archives/income_wealth/cb10–144.html (accessed September 27, 2010).
2. *National Geographic,* insert of July 2010 issue.
3. Blog Pulse, "BlogPulse Stats for September 21, 2010," http://www.blogpulse.com/ (accessed September 21, 2010). See also Anne Helmond, "How Many Blogs Are There? Is Someone Still Counting?" *The Blog Herald,* February 11, 2008, http://www.blogherald.com/2008/02/11/how-many-blogs-are-there-is-someone-still-counting/ (accessed September 21, 2010). See also Gary M. Stern, "Keeping Track of the Ever-Proliferating Number of Blogs," Information Today, Inc., Feb 15, 2010, http://www.infotoday.com/linkup/lud021510-stern.shtml (accessed September 21, 2010).
4. Robert Bacal, "Are You Building An Online Group, Community, Or Pseudo-Community?" June 21, 2010, http://www.customerthink.com/blog/are_you_building_an_online_group_community_or_pseudo_community (accessed September 11, 2010).
5. James Poniewozik, "The Year in Culture: Has the Mainstream Run Dry?" *Time* magazine, December 29, 2003, 149.
6. Brooks Barnes, "Guardians of TV Morals, Bowed," *New York Times,* October 25, 2010, B1.
7. Kwame Anthony Appiah, "The Case for Contamination," *New York Times Sunday Magazine,* January 1, 2006, sec. 6, 30–35, 51–52.
8. John Harwood, "If Fox Is Partisan, It Is Not Alone," *New York Times,* November 2, 2009, A12.
9. James Davison Hunter, *Culture Wars: The Struggle to Define America* (New York: Basic Books, 2002), 49–50.
10. David Lowenthal, *The Past Is a Foreign Country* (New York: Cambridge University Press, 1985), 197.
11. Mark Bauerlein, *The Dumbest Generation: How the Digital Age Stupefies Young Americans and Jeopardizes Our Future* (New York: Tarcher/Penguin, 2008), 8.

12. E. D. Hirsch, *Cultural Literacy: What Every American Needs to Know* (New York: Vintage, 1988), 10 & 29.
13. Alvin Kernan, *In Plato's Cave* (New Haven, CT: Yale University Press, 1999), 232.
14. Hunter, *Culture Wars,* 57–59.
15. Ralph Keyes, *The Post-Truth Era: Dishonesty and Deception in Contemporary Life* (New York: St. Martin's Press, 2004). See also Farhad Manjoo, *True Enough: Learning to Live in a Post-Fact Society* (Hoboken, NJ: Wiley, 2008).
16. Susan Jacoby, *The Age of American Unreason* (New York: Pantheon Books, 2008), 6.
17. James Poniewozik, "The Myth of Fact," *Time* magazine, August 23, 2010, 62.
18. FactCheck.org, "Republican-Funded Group Attacks Kerry's War Record," August 6, 2004, updated August 22, 2004, http://www.factcheck.org/article231.html (accessed September 24, 2010).
19. Angie Drobnic Holan, "PolitiFact's Lie of the Year: 'Death Panels,'" Friday, December 18, 2009, http://politifact.com/truth-o-meter/article/2009/dec/18/politifact-lie-year-death-panels/ (accessed September 23, 2010).
20. Brian Stelter, "Debunkers of Fictions Sift the Net," *New York Times,* April 5, 2010, B1 & B2.
21. FactCheck.org, http://www.factcheck.org/about/ (accessed September 23, 2010).
22. Andrew Keen, *The Cult of the Amateur: How Today's Internet Is Killing Our Culture* (New York: Doubleday, 2007), 42, 186. See also Alvin Kernan, *In Plato's Cave,* 244, and Richard Hofstadter, *Anti-Intellectualism in American Life* (New York: Alfred Knopf, 1963), 154.
23. Virginia Heffernan, "Pop-up Stars: How Reality TV Enables Authentic Phoniness," *New York Times Sunday Magazine,* June 13, 2010, 24–26.
24. Stephanie Clifford, "In an Era of Cheap Photography, the Professional Eye is Faltering," *New York Times,* March 31, 2010, B1 & B5.

25. National Football League, http://fantasy.nfl.com/registration/league Directory (accessed September 17, 2010).
26. http://technorati.com/blogs/directory/ (accessed September 14, 2010).
27. Alexa the Web Information Company, "Estimate that about 25% of global internet users . . ." http://www.alexa.com/siteinfo/youtube.com (accessed September 14, 2010). Also Answers.com, "How many videos are there on YouTube?" http://wiki.answers.com/Q/How_many_videos_are_there_on_YouTube (accessed September 14, 2010); Artem Russakovskii, "How to Find Out the Number of Videos on YouTube," August 14, 2008, updated August 8, 2010, http://beerpla.net/2008/08/14/how-to-find-out-the-number-of-videos-on-youtube/ (accessed September 14, 2010); Michael Arrington, TechCrunch, "YouTube Video Streams Top 1.2 Billion/Day," June 9, 2009, http://techcrunch.com/2009/06/09/youtube-video-streams-top-1-billion day/ (accessed September 14, 2010).
28. Kalle Lasn, *Culture Jam: How to Reverse America's Suicidal Consumer Binge* (New York: Quill, 1999), 11.
29. Dave Itzkoff, "The Refined Art of Tastelessness," *New York Times Sunday Week in Review,* August 8, 2008, 2.
30. Alastair Macaulay, "This Time the Trouble Isn't Wicked Step Sisters," *New York Times,* September 6, C1 & C6.
31. Nancy Franklin, "Talking Dirty: Chelsea Handler Sexes up Late Night," *New Yorker,* May 24, 2010, 68–70.
32. On Andrew Breitbart's use of such words as cunt and cocksucker, see Rebecca Mead, "Rage Machine: Andrew Breitbart's Empire of Bluster," *New Yorker,* May 24, 2010, 26–32.
33. Jacoby, 7.
34. Stuart Elliott, "Bleep or No Bleep, Bolder Words Blow In," *New York Times,* May 14, 2010, B1 & B4.
35. Elliott, "Bleep or No Bleep," B4.
36. Janet Abbate, "Privatizing the Internet: Competing Visions and Chaotic Events," *Annals of the History of Computing, IEEE* 32, no. 1 (2010): 10–22, http://ieeexplore.ieee.org/xpl/freeabs_all.jsp?arnumber=5430757 (accessed September 1, 2010).

37. See Chong Fu and Hui-yan Jiang, "Privatizing the Internet: Competing Visions and Chaotic Events," 2008 International Conference on Intelligent Computation Technology and Automation (ICICTA), October 20–22, 2008, 840–44. posted October 28, 2008, http://ieeexplore.ieee.org/xpl/mostRecentIssue.jsp?punumber=4659422 (accessed September 23, 2010). Also J. C. Sprott, "Numerical Calculation of Largest Lyapunov Exponent," revised August 31, 2004, http://sprott.physics.wisc.edu/chaos/lyapexp.htm (accessed September 23, 2010).
38. Todd Defren, Pr-Squared, "Cultural Fragmentation, the Long Tail, & Your 15 Minutes," http://www.pr-squared.com/index.php/2005/10/cultural_fragmentation_the_lon (accessed September 9, 2010).
39. Cass Sunsten, "Is the Internet Really a Blessing for Democracy?" *The Daily We, Boston Review,* Summer 2001, http://bostonreview.net/BR26.3/sunstein.php (accessed September 10, 2010).
40. John Early McIntyre, "Allude at Your Own Risk," June 22, 2005, http: //www.poynter.org/content/content_view.asp?id=8350 (accessed September 10, 2010).
41. Keen, *The Cult of the Amateur,* 2–3.
42. Keen, *The Cult of the Amateur,* 102.
43. Marketing Shift Online, "Tower Collapse a Warning," August 21, 2006, http://www.marketingshift.com/2006/8/tower-collapse-a-warning.cfm (accessed September 18, 2010).
44. David Goldman, "Music's Lost Decade: Sales Cut in Half," CNN.com, February 3, 2010, http://money.cnn.com/2010/02/02/news/companies/napster_music_industry/ (accessed September 21, 2010).
45. Bram Teitelman, "Music Industry Decline Goes Beyond Recorded Music," May 4, 2010, Metal Insider, http://www.metalinsider.net/gloom-and-doom/music-industry-decline-goes-beyond-recorded-music (accessed September 15, 2010).
46. Goldman, "Music's Lost Decade."
47. Claire Suddath, "Rock Steady," *Time* magazine, August 9, 2010, 51–53.
48. Daniel J. Wakin, "Philharmonic President Is to Depart, as Music World Changes," *New York Times,* September 28, 2010, C6.

49. Wakin, "Philharmonic," C1.
50. See wordsong.org.
51. Joseph A. Schumpeter, *Capitalism, Socialism and Democracy* (New York: Harper, 1975, originally published 1942), 82–85.
52. Michael J. de la Merced, "Blockbuster, Hoping to Reinvent Itself, Files for Bankruptcy," *New York Times,* September 24, 2010, B3.
53. David Gelernter, *1939: The Lost World of the Fair* (New York: Free Press, 1995), 42.
54. With the loss of perhaps $6 trillion from the stock market and perhaps $1 trillion from the housing market, little wonder that people worry. See Gail Marksjarvis, "The Stock Market Honeymoon Ends as Bitter Losses Remain," *Chicago Tribune,* August 13, 2010, http://newsblogs.chicagotribune.com/marksjarvis_on_money/2010/08/the-stock-market-honeymoon-ends-as-bitter-losses-remain.html (accessed September 14, 2010). Also see Lorraine Woellert and John Gittelsohn, "Fannie-Freddie Fix at $160 Billion with $1 Trillion Worst Case," *Bloomberg Business Week,* June 14, 2010, http://www.businessweek.com/news/2010–06–14/fannie-freddie-fix-at-160-billion-with-1-trillion-worst-case.html (accessed October 2, 2010).
55. Emily Steel and Jessica E. Vascellaro, "Facebook, MySpace Confront Privacy Loophole," *Wall Street Journal,* May 21, 2010, http://online.wsj.com/article/SB10001424052748704513104575256701215465596.html (accessed Oct 6, 2010).
56. Richard D. Smith, "Responding to Global Infectious Disease Outbreaks: Lessons from SARS on the Role of Risk Perception, Communication and Management," *Social Science and Medicine* 63, vol. 12 (2006): 3113–123, doi:10.1016/j.socscimed.2006.08.004. PMID 16978751, http://www.sciencedirect.com/science?_ob=ArticleURL&_udi=B6VBF-4KWTFM1–2&_user=10&_coverDate=12%2F31%2F2006&_rdoc=1&_fmt=high&_orig=search&_origin=search&_sort=d&_docanchor=&view=c&_acct=C000050221&_version=1&_urlVersion=0&_userid=10&md5=1e6cdc185de938c209c47f295b6cca27&searchtype=a (accessed November 1, 2010).
57. Richard Preston, *Panic in Level 4* (New York: Random House, 2008), 55.

58. New York Civil Liberties Union, *A Special Report: Who's Watching? Video Camera Surveillance in New York City,* Fall, 2006, http://www.nyclu.org/pdfs/surveillance_cams_report_121306.pdf (accessed September 12, 2010).
59. Edward J. Blakely and Mary Gain Snyder, *Fortress America: Gated Communities in the United States* (Cambridge, MA: Brookings Institute and Lincoln Institute, 1999), 2.
60. Blakely and Snyder, *Fortress America,* 1.
61. Hava El Nasser, "Gated Communities More Popular, and Not Just for the Rich," *USAToday,* December 12, 2002, posted 12/16/2002, http://www.usatoday.com/news/nation/2002–12–15-gated-usat_x.htm (accessed September 10, 2010).
62. Blakely and Snyder, *Fortress America,* 1.
63. Gelernter, *1939,* 28.
64. Michael St. Clair, *Millenarian Movements in Historical Context* (New York: Garland Publishing, 1992), ch. 1.
65. Keen, *The Cult of the Amateur,* 113.
66. Nick Georgano (ed.), *The Beaulieu Encyclopedia of the Automobile* (Chicago: Fitzroy Dearborn, 2000).

Chapter 9

1. Ann Kirschner, "My iPad Day," *The Chronicle Review,* June 13, 2010, http://chronicle.com/article/My-iPad-Day/65839/ (accessed July 5, 2010).
2. Virginia Heffernan, "Sweetness and Backlight," *New York Times Sunday Magazine,* July 4, 2010, 14–15.
3. Andrea B. Bear and Kathleen A. Brehony, "Changing How Organizations Manage Change from the Inside Out," in *Changing the Way We Manage Change,* ed. Ronald R. Sims (Westport, CT: Quorum Books, 2002), 224.
4. Bear and Brehony, "Changing," 20.
5. Ronald R. Sims, "General Introduction and Overview of the Book," in *Changing the Way We Manage Change,* ed. Ronald R. Sims (Westport, CT: Quorum Books, 2002), 1.

6. Otto Friedrich, *Decline and Fall: The Struggle for Power at a Great American Magazine* (New York: Harper & Row, 1970), 13.
7. Valorie Burton, *How Did I Get So Busy? The 28-Day Plan to Free Your Time, Reclaim Your Schedule, and Reconnect with What Matters Most* (New York: Broadway Books, 2007), 2, 9, 13.
8. Beth Sawi, *Coming Up for Air: How to Build A Balanced Life in a Workaholic World* (New York: Hyperion, 2000), 8.
9. William Powers, *Hamlet's Blackberry: A Practical Philosophy for Building a Good Life in the Digital Age* (New York: HarperCollins, 2010), 218.
10. C. M. Boots-Faubert, "Surviving a Week Without Technology," *Cape Cod Times,* August 31, 2010, C1.
11. Boots-Faubert, C2.
12. Susan Dominus, "Encouraging the Text Generation to Rediscover Its Voice," *New York Times,* April 27, 2010, A16.
13. Powers, 215–19.
14. Carl R. Sunstein, *Going to Extremes: How Like Minds Unite and Divide* (New York: Oxford University Press, 2009), 144.
15. Sunstein, *Going to Extremes,* 147.
16. Kwame Anthony Appiah, "The Case for Contamination," *New York Times Sunday Magazine,* January 1, 2006, 30–35, 51–52.
17. Jonathan K. Pritchard, "How We Are Evolving," *Scientific American,* October, 2010, 41.
18. Pritchard, 45–46.
19. Mark C. Taylor, *The Moment of Complexity: Emerging Network Culture* (Chicago: University of Chicago Press, 2003), 19.
20. Taylor, *The Moment of Complexity,* 19.
21. Taylor, *The Moment of Complexity,* 13.
22. Joseph A. Tainter, *The Collapse of Complex Societies* (Cambridge: Cambridge University Press, 1990).
23. Jared Diamond, *Collapse: How Societies Choose to Fail or Succeed* (New York: Penguin Books, 2005).

24. Thomas L. Friedman, "Build 'Em and They'll Come," *New York Times,* October 13, 2010, A23.
25. Douglas Rushkoff, *Playing the Future: How Kids' Culture Can Teach Us To Thrive in an Age of Chaos* (New York: HarperCollins, 1996), 17.
26. Freeman Dyson, "What Price Glory," review of *Lake Views: This World and the Universe,* by Steven Weinberg, *New York Review of Books,* June 10, 2010, 8–12.
27. Mark Elvin, *The Retreat of the Elephants: An Environmental History of China* (New Haven: Yale University Press, 2004), 123 & 492.
28. Joe Sharkey, "Reinventing the Suitcase by Adding the Wheel," *New York Times,* October 5, 2010, B6.
29. Robin Marantz Henig, "In Vitro Revelation," *New York Times,* October 5, 2010, A27.
30. Pete Hamill, *A Drinking Life: A Memoir* (Boston: Little, Brown, 1994), 24.
31. Bill Bryson, *The Life of the Thunderbolt Kid: A Memoir* (New York: Broadway Books, 2006), 5.
32. Bryson, *Thunderbolt Kid,* 6.
33. Bryson, *Thunderbolt Kid,* 9.
34. Bryson, *Thunderbolt Kid,* 11.
35. Liaquat Ahamed, *Lords of Finance: The Bankers Who Broke the World* (New York: Penguin Press, 2009), 9.
36. Stefan Zweig, *The World of Yesterday: An Autobiography,* trans. Helmut Ripperger (New York: Viking Press, 1943), vii.
37. Alvin Kernan, *In Plato's Cave* (New Haven: Yale University Press, 1999).
38. Wendy Steiner, *Venus in Exile: The Rejection of Beauty in Twentieth-Century Art* (New York: Free Press, 2001); and Martha Bayles, *Hole in Our Soul: The Loss of Beauty and Meaning in American Popular Music* (New York: Free Press, 1994).

Index

About the Author

Michael St. Clair is Professor of Psychology, Emeritus, at Emmanuel College, Boston, and has taught courses on extremism, the future, Byzantine history, and various areas of psychology, at both the undergraduate and graduate level. He holds master's degrees in classics, philosophy, and religious studies. He received his PhD from Boston University and worked as a licensed clinical psychologist for 20 years, doing marriage and family therapy. His research increasingly is in the area of tracking and measuring change. He has written several books and articles, among them: *Object Relations and Self Psychology: An Introduction* and *Millenarian Movements in Historical Context.*